Robot Development Using Microsoft® Robotics Developer Studio

Shih-Chung Kang • Wei-Tze Chang
Kai-Yuan Gu • Hung-Lin Chi

CRC Press
Taylor & Francis Group
Boca Raton London New York

CRC Press is an imprint of the
Taylor & Francis Group, an **informa** business

A CHAPMAN & HALL BOOK

CRC Press
Taylor & Francis Group
6000 Broken Sound Parkway NW, Suite 300
Boca Raton, FL 33487-2742

© 2011 by Taylor & Francis Group, LLC
CRC Press is an imprint of Taylor & Francis Group, an Informa business

No claim to original U.S. Government works

Printed in the United States of America on acid-free paper
Version Date: 20110512

International Standard Book Number: 978-1-4398-2165-7 (Hardback)

Visit the Taylor & Francis Web site at
http://www.taylorandfrancis.com

and the CRC Press Web site at
http://www.crcpress.com

Table of Contents

Preface

This book has evolved from class notes for the undergraduate Automation and Robotics course at the National Taiwan University. The course was originally designed for senior students in the Department of Civil Engineering, enhancing their understanding of robotics technology and its potential. As this course does not have any prerequisites, it has been adopted as a fundamental robotics class by many colleges in Taiwan.

In my experience with robot education, a top-down teaching approach works better than teaching from fundamental theories. A top-down approach can help students think of robots as a system rather than an assemblage of parts. Students will gain an understanding of methods for integration, design trade-offs, and even teamwork, which are essential skills to building robots. I have therefore designed many smaller examples, one for each important concept in the course. Students work in groups and cooperate to follow course notes and to implement each example, and then try to decipher the mechanism and theories behind the examples. I have found that the top-down approach is very effective and keeps students learning actively.

For the course, I selected Microsoft Robotics Developer Studio (MSRDS) as the software platform and the Lego NXT education kit as the hardware platform. This combination is ideal. MSRDS has modularized the sensors and actuators as different software services. Students can use a visual programming language to link the services as visual information flows between services. Programming the visual language is straightforward and even students with little programming experience can learn it in a short time, given that they are familiar with the logic of robot controls.

The Lego NXT Education kit (#9797) is an excellent teaching aid for three reasons. First, it contains a variety of sensors. The kit includes four types of common sensors: touch, ultrasonic, light, and sound. Students have ample opportunities to learn the qualities of the different sensors and understand how to use them correctly and cleverly. Second, the kit has excellent compatibility with MSRDS. MSRDS provides software services for all the sensors and actuators, and thus students need not deal with tedious communication protocols between computers and robot components. If, for example, a sensor is added to the robot, the students need only drag and drop a sensor icon into the programming environment, from which the sensor readings can be taken immediately. The third reason is that

Lego has a long history in the toy industry and has developed a great variety of brick shapes with which students are able to compose any form of robot they can imagine.

This book includes an extensive array of examples, with corresponding step-by-step tutorials and explanations for use by novice robot developers. Readers may learn at their own pace following these examples. Alternatively, instructors may choose this book as reference for a one-semester course. Instructors should emphasize the main concepts of the examples and allow students to practice the examples by themselves. The book is organized into ten chapters as follows.

CHAPTER 1: CHALLENGES AND SOLUTIONS IN ROBOTICS

Along with the recent rapid progress in information and electromechanical technology, research in service-type robots and their applications has become easier to perform and increasingly widespread. Robotic research is now a well-explored field of study, and it is possible that service-type robots will become the key enablers for another industrial revolution. This chapter compares the technological aspects of both traditional industrial robots and service-type robots and discusses the research and development challenges for service-type robots. It also recommends the scaffolding teaching method for adopting a practical approach toward learning about service-type robots.

CHAPTER 2: MICROSOFT ROBOTICS DEVELOPMENT PLATFORM

The control program for the service-type robot requires a distributable and highly scalable software platform. This chapter introduces the Microsoft Robotics Developer Studio (MSRDS) to target those requirements and to make it easier for experts of different domains to conduct robotic research and development. This chapter introduces the development origins of MSRDS and details the relevance of the core technology used in service-type robot software. The necessary software and hardware environment and installation steps for MSRDS installation are also detailed.

CHAPTER 3: MICROSOFT VISUAL PROGRAMMING LANGUAGE

MSRDS provides an intuitive program development environment—Microsoft's Visual Programming Language (MVPL). MVPL uses a visual interface, which is extremely effective for displaying a service-type robot's software logic and reasoning. Inexperienced programmers can easily adapt themselves to service-type robot program design through use of MVPL, while expert programmers can use MVPL to display complex logic and specialized algorithms. This chapter uses six examples to explain MVPL's program design logic, enabling readers to quickly learn MVPL operations.

CHAPTER 4: VISUAL SIMULATION ENVIRONMENT

During robotic research and development, hardware costs are typically high. MSRDS addresses this issue by providing the Visual Simulation Environment (VSE), which allows robot developers to test robot behaviors in a simulated environment. This is advantageous for service-type robot research groups and for novice programmers alike. This chapter uses six examples to describe the robot programming process under VSE, enabling readers to understand MSRDS's simulation environment.

CHAPTER 5: ROBOT I/O UNIT

Starting from this chapter, the book formally describes actual robot operations. This book uses the LEGO MINDSTORMS Education NXT Base Set Robots to introduce commonly seen robot component compositions, including a wireless Bluetooth module, various sensors (e.g., touch, light, sound, and ultrasonic), and transmission motors. By using real case studies, we guide the reader to understanding the various input and output components and the interoperability between components.

CHAPTER 6: ROBOT MOTION BEHAVIOR

This chapter introduces MSRDS's hardware manifest. This design can store preassembled hardware setups, which reduces the time and complexity for developing service-type robot control programs. Further to this, Chapter 6 describes how to couple sensor components and transmission motors to design an autonomous robot's operational behavior. From this chapter, the reader can gain a working knowledge of the properties and usage characteristics of each sensor device.

CHAPTER 7: CONTROLLING THE ROBOT THROUGH SOUNDS

Human–robot interaction is an important topic for service-type robots. This chapter introduces the voice control functionality in MSRDS. Through voice recognition technology, be it the voice of the elderly or of children, one can easily control a service-type robot's behavior. This application extends into industries such as home care and entertainment, allowing the service-type robot to find application in modern family life. This chapter introduces MSRDS's built-in voice recognition service and application possibilities.

CHAPTER 8: ROBOT VISION

Image processing technology is reaching maturity and service-type robots can use this as leverage for addressing the inadequacies of other sensor components. Although the LEGO Education robot does not provide visual components, MSRDS allows the use of web cameras to develop the robot's visual system. This chapter couples the LEGO Education robot with the web camera and uses it to demonstrate the setup of the visual system and its operation.

CHAPTER 9: A REAL APPLICATION—SUMO ROBOT CONTEST

A great way to learn about the service-type robot is through a contest. This chapter introduces the design of a sumo robot. Both the preparation and contest results provide a real learning experience. The contest also serves as reference for teaching about service-type robots. From this chapter, readers can also come to understand varied and flexible designs for a service-type robot.

CHAPTER 10: RELATED LEARNING RESOURCES

This chapter introduces related learning material for MSRDS and the LEGO Education robot, which includes course websites from other institutions, the MSRDS and LEGO Education robot's web forum, and other education websites. This helps readers deepen their understanding of service-type robots and aids their learning.

Acknowledgments

This book could not have been published without the support of many people. I would first like to recognize the contributions of the coauthors, Wei-Tze Chang, Kai-Yuan Gu, and Hung-Lin Chi. They worked with me closely over the past year to make this book as close to perfect as possible. I also would like to express my appreciation for the support from my colleagues at National Taiwan University (NTU), especially Professor Patrick Hsieh and Professor David Chen. Their support provided me with an opportunity to consolidate my teaching material into a book. I would like to acknowledge the advice and encouragement from several robot experts: Tandy Trower (Microsoft), Young Joon Kim (Microsoft), Professor Jean-Claude Latombe (Stanford University), and Professor Ren Luo (NTU). Their words deeply inspired me to devote more of myself to robot education and applications. I would like to thank Taiwan's National Science Council and the Department of Civil Engineering at NTU for their generous grants. Thanks also to Uni-edit for their excellent English translation of the book.

Finally, I appreciate the patience and sacrifices of my family over the past year while I wrote this book. Thank you to my wife, Lily Chang, and my sons, David Kang and Joshua Kang.

Shih-Chung Jessy Kang
National Taiwan University
Taipei, Taiwan

Authors

Shih-Chung (Jessy) Kang, PhD, holds a doctorate degree from Stanford University and is currently an associate professor with the Department of Civil Engineering at National Taiwan University (NTU). He began using MSRDS on a research project in early 2006, half a year before its official launch by Microsoft. In 2008, he started offering a hands-on undergraduate course for students to learn about robot development. In the same year, he earned a teaching excellence award from NTU. In 2009, he started providing training courses in Taiwan to share his teaching experiences with other instructors in the field of robotics.

Wei-Tze (Aries) Chang, PhD, earned his doctorate degree from National Taiwan University (NTU) and is currently a postdoctoral researcher there. He is also an adjunct assistant professor with the Department of Civil Engineering at Tamkang University in Taiwan. He has been working with engineering software development and high-performance computing techniques for over 5 years. Recently his interests have expanded into the field of robotics education, bringing him into Dr. Kang's group. Currently, he is responsible for the training of teacher leaders in service-type robots for the Robotics Society of Taiwan.

Hung-Lin Chi is currently a doctoral student in the Department of Civil Engineering at NTU. While studying for his master's degree, he focused on applying virtual techniques to construction machines, which included the simulation of the physical behaviors of cranes, the analysis of crane operations, and the development of a motion planning system. Currently his research interest is the integration of robot-sensing techniques into remote-controlled systems for cranes.

Kai-Yuan Gu, an NTU master's graduate, currently works as a digital design engineer with the Mighty Power Solutions Corp. While at NTU, he focused on the application of robotics in civil engineering, which included the design of an autonomous robot for pavement inspections. Currently he is designing automated products for energy-saving devices and lighting.

CHAPTER 1

Challenges and Solutions in Robotics

THE TERM *ROBOT* ORIGINATES FROM the fictional novel titled *Rossum's Universal Robots*˙ published in 1921 by the Czechoslovakian writer Karel Capek. In the novel, an R.U.R. (i.e., Rossum's Universal Robot), which is the equivalent of a robot, is described to be composed of a collection of components and is smart enough to replace the humans working in a designed area. Soon after Capek's novel was released, the concept of robots was illustrated in theaters and movies, stimulating the public's interest in the emergence of robots and their development and encouraging further research into this field.

Nevertheless, in the real world, it is not easy to create a smart robot as envisioned in Karel Capek's novel. A smart robot is an entity that can operate autonomously and combines technological breakthroughs in machinery, automation, electrical, optics, electronics, software, communication, safety mechanisms, and creative design. It is a product that merges high-end technologies and has a high commercial value. As illustrated in Figure 1.1, knowledge about robot development crosses several domains and can be broadly classified into four layers, namely hardware, software, applications, and user layers.

1. Hardware layer: This determines the hardware capabilities of the robot, which is analogous to the bodily functions of a human being. This layer includes the manipulators, sensors, actuators, controllers, and so on. Hardware construction requires expert knowledge about machinery, automation, electrics, optics, and electronics.

2. Software layer: This determines the responsiveness of the robot, which is analogous to the senses and reactions of a human being. The categories in this layer include subjects such as concurrent programming and the robot's perception and simulations. It requires software and communication protocols for construction.

˙   Capek, Karel (2001). R.U.R., translated by Paul Selver and Nigel Playfair, Dover Publications, p. 49.

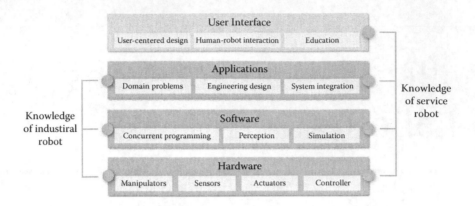

FIGURE 1.1   The four layers of knowledge in robot development.

3. Application layer: This determines the usage of the robot and the application domain, which is analogous to a human being's specialization and profession. This layer concerns domain problems, engineering design, and system integration. As a result, during construction of the robot, participation is required from engineers and experts of various domains.

4. User interface layer: The first three layers—that is, the hardware, software, and application layers—form the main components of the industrial robot. However, as robotics advances technologically and applications spread, service robots, which is a new type of robot, cannot be built using knowledge from the three previous layers alone. Service robots require the inclusion of the user interface layer. This layer embeds a user-centered design (UCD), human-robot interaction (HRI), robot education, and other similar topics. Construction of this layer requires knowledge from the pedagogical, psychological, and scientific fields.

Due to the fact that robotics crosses multiple knowledge domains, difficulty and complexity are introduced into its research and development. The following sections discuss the challenges faced from the perspective of smart-robot development.

1.1   RESEARCH CHALLENGES: MOVING FROM INDUSTRIAL-TYPE TO SERVICE-TYPE ROBOTS

As mentioned in the previous section, we can classify robots based on their functionalities into industrial-type robots and service-type robots. In Figure 1.2, the left image represents a typical industrial robot and the right image represents a service robot.

Industrial robots are focused mainly on manufacturing. As their deployed environment is known and basic, the robots are usually fixed at a certain location. An industrial robot's design goal is focused on precision control, rather than on goals such as artificial intelligence. Consequently, these robots do not require large amounts of computational ability, although they are considered the most mature type of robot technology available, with very stable and reliable system functionalities. At present, industrial robots are automated to the point that they replace the work of humans in performing

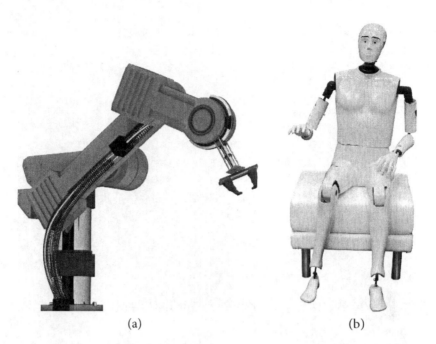

(a) (b)

FIGURE 1.2 Robots classified as industrial-type and service-type based on their functionalities. (a) Industrial-type robots: single purpose, sequential procedure, goal-oriented, known environment, high precision. (b) Service-type robots: multiple purpose, concurrent procedure, task-oriented, uncertain environment, fuzzy decision-making.

tasks such as component assembly, processing, and welding. Industrial robots have been widely used in industries such as automobile manufacturing and precision electronics manufacturing.

In contrast to industrial robots, the goal of a service robot is to provide a service to address the needs of humans. Being closer to human habitats, the environment to which service robots are deployed is unknown and there is a need for reactions to sudden changes. In terms of executing tasks, the industrial robot need only repeat the execution of a single task, whereas service robots may need to execute a range of tasks and perform decision making in unknown situations. As a consequence, the design of a service robot is much more complicated than that of an industrial robot in terms of both hardware and software. For instance, service robots require a highly flexible hardware construction, a flexible design, multitasking capabilities, and the capability to make adjustments and corrections quickly and easily. In addition, it is necessary for service robots to appear sincere and friendly, as they will interact with humans directly.

Since the 1960s, an increasing number of domain experts have invested time and effort in robotic research work. However, during those earlier years, computers had very limited computational capabilities and this limited the growth of associated robotic research areas such as artificial intelligence. Industrial robots were the main driving force behind robotic research during these early stages. In the past two decades, significant advances have been made to computational capabilities and robotics technology has progressed rapidly from experimental forms to real-life applications. Robots, like humans, are now able to perform actions independently. As such, service robots that can autonomously make decisions have attracted both research and industrial interests. Their possible applications are actively being explored and discovered.

1.2 CHALLENGES IN THE SOFTWARE PLATFORM FOR SERVICE ROBOTS

Technological hardware advances made over the past decades have given consumers the flexibility of choice in hardware components. These hardware advances are the result of research in building precise sensing instruments, stable control devices, and highly efficient transmission motors. However, although these hardware improvements have vastly extended robots' physical functionalities, the resulting performance is still constrained by the operating software. Currently, software development on robots is faced with three research challenges:

1. How can the software coordinate multiple tasks?

2. How can the software process tasks and events in sync?

3. How can we unify the development platform?

1.2.1 Challenge 1: How Can the Software Coordinate Multiple Tasks?

A typical electronic appliance or equipment has only a single purpose—for example, an automatic door is only required to open or shut and a fan only needs to spin to generate wind. In comparison, an intelligent robot differs in that its design is to support multiple different tasks to fulfill its goal of performing numerous complex duties. For example, a robot that faces an obstacle while moving toward a target destination should be able to avoid the obstacle and continue proceeding toward the destination without human intervention. Although this seems to be a basic action that even a young child can accomplish, for a robot (or should one say, for the robot designers) it is a tedious and complex procedure. More specifically, the robot in this situation is similar to the toddler: it needs to distinguish first what is considered an obstacle, determine the dimensions of the obstacle, and perform some calculations. According to its own capabilities, experiences, and knowledge of avoidance plans from a repository, it then executes the avoidance task. Here, the *avoidance object* and *target destination* are the two goals that the robot has to simultaneously consider while moving. This example illustrates the significance of the research problem of coordinating multiple duties. Nevertheless, this is a very basic demonstrative example. Instead, if the robot was assigned to a more complex job such as aged care or security patrolling, the required duties would be more varied and the behavior design of these robots would be significantly more difficult.

1.2.2 Challenge 2: How Can the Software Process Tasks and Events in Sync?

A human conceptually constructs the place he or she is in through the information derived from the five senses, namely sight, hearing, smell, taste, and touch. Similarly, a robot utilizes sensors with differing capabilities to collect data about its environment before deducing an appropriate reaction. However, the data provided by sensors can arrive asynchronously in a complex and rapidly changing manner. Therefore, for robot software design, it is a challenge to combine all the data received while actively controlling the robot's motion. This is a limitation of the conventional approach of robot design to utilize sequential procedure software that enables a robot to perform only one task at a time. Thus the software cannot process sensed readings while also controlling the motion element. As a result, it is unable to react to changes swiftly. As an example, consider how we normally cross a road. To cross a road, we simultaneously observe traffic coming from our left and right (i.e., sensing) and quickly cross the road when it is clear (i.e., movement). These two tasks occur simultaneously and do not interfere with each another, or in other words, they are performed in a concurrent procedure. Although the concurrent procedure concept, commonly referred to as parallel computing, distributed computing, and multi-thread processing, is well developed, its applications in robots are still rudimentary. In particular, it is very difficult for a concurrent program to extract the essence of sensed information (e.g., determining incoming vehicles) and, at the same time, control movement (e.g., crossing the road). Issues such as determining whether or not to sense and move a robot simultaneously, prioritizing the tasks to perform, and dissecting and processing information at the same time collectively present a major challenge to building a robot's program design.

1.2.3 Challenge 3: How Can We Unify the Development Platform?

As early industrial robots incorporated only basic functionalities, the design of each robot was targeted to meet industry demands and components. Consequently, different software systems were developed, such as the ERSP platform originating from the Evolution Robotics and the commercial platform NI as proposed by National Instruments (NI). There are also some other platforms for research purposes, such as the Player/Stage platform originating from the National Science Foundation (NSF) and Stanford University. Although these software systems are incompatible with one another, each system performs well in its respective area. However, this incompatibility is a software barrier in the development of universal robots. In addition, it would be unfortunate if the robot software developed for one system could not be easily ported for use in another system. The list of issues to be addressed includes designing a single common interface for universal robots, enabling a central public repository for robot software, giving developers the flexibility to reuse previously implemented software components, and so on. All of these issues represent major challenges for software platform design.

1.3 EDUCATIONAL CHALLENGES IN THE DEVELOPMENT OF SERVICE ROBOTS

The successful development of robotics research is heavily reliant on contributions from domain experts, similar to many other fields of study. At the present stage of robotics education, courses relating to the hardware layer (such as hardware organization design, machine dynamics, and electronics) and software design (such as parallel processing, image recognition, and artificial intelligence) have developed progressively over the years and are offered by various universities, colleges, and institutions. However, robotics is an applied science and when there is insufficient emphasis on the education of its applications, robotics education will not be sufficiently thorough. This issue is currently reflected in the current robots available on the market; while they embody state-of-the-art technologies, they often cannot be used to solve practical problems. This problem would be resolved if more application designers were involved in the development of robots. For instance, robots may be useful in civil engineering projects that are sometimes high-risk and repetitive, such as underground line channel exploration, tunnel excavation, and bridge and road inspections. If smart robots can assist in the execution of such tasks, it would not only reduce the number of accidents on projects but also improve the overall quality of the project outputs. Moreover, in the case of civil engineering, a smart robot would encompass target goals in its design phase such as resistance to vibration, dust, and water, as well as being able to walk on rugged road surfaces.

However, the main challenge remains in addressing the different levels of domain expertise regarding robots, resulting in communication difficulties when applying contributors' ideas to the design of a particular robot. Such difficulties negatively affect the development of robots. Furthermore, due to the vast library of knowledge in robotic studies, which spans multiple disciplines, the present form of education in robotics (from components to system integration) does not adequately prepare the domain experts. Therefore it remains a major challenge for education in robotics to sufficiently educate and train domain experts.

1.4 TOP-DOWN LEARNING STRATEGY

The aforementioned three categories of challenges—education, research, and software—are the current bottlenecks in the development of robots. Although the first two challenges (i.e., how to develop intelligent robots and how to create an integrated software development platform) remain controversial issues, we are able to observe trends in research and software development and can, in the meantime, address educational challenges that form the basis of robotics knowledge. In the four layers of the robotics knowledge structure, relevant courses are already quite well developed for the two lower layers, that is, the hardware and software layers. However, the same cannot be said for the upper two layers, that is, the application and user interface layers. Upon closer examination, this is because the upper two layers involve a large knowledge base that is complex in nature and vastly different from the knowledge of the lower two layers (typically learned in automatic control, mechanical, electronic, electrical, and software engineering courses). Hence it is impractical to design courses that address the knowledge required in all four layers. It is necessary to construct an education program that can focus on the skills required for the upper two layers, so that those who are already equipped with knowledge of the two upper layers can quickly absorb necessary knowledge of the lower layers and thus be prepared for development in the field of robotics.

We suggest using *scaffolding instruction* to design a rapid teaching technique that surpasses software and hardware knowledge layers. This is a top-down teaching strategy as illustrated in Figure 1.3. In this form of teaching technique, as the target audiences are

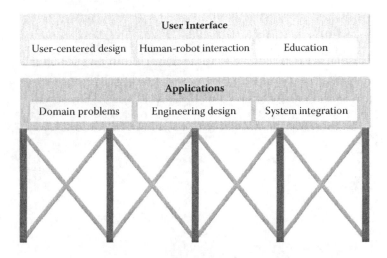

FIGURE 1.3 Scaffolding instruction as applied to robotics education.

interested in the application of the robots, a problem-based approach is more appropriate for teaching students how to use robots to complete certain tasks in an application scenario. In this book, the scaffold for the software and hardware layers is constructed using teaching aids, namely the LEGO Mindstorm Education NXT Base Set robots (henceforth abbreviated as "LEGO robots") and the Microsoft Robotics Developer Studio (MSRDS). This allows students to directly target application-specific problems to learn robotics. In the future, if the students develop an interest in the software and hardware layers, they can use LEGO robots and MSRDS or even a different scaffold to further improve their hardware skills, software skills, or both.

1.5 USING LEGO MINDSTORM EDUCATION NXT BASE SET ROBOTS AND MSRDS

From our teaching experiences, we found that LEGO robots and MSRDS can be an ideal combination to support the education of service-type robots. This combination provides excellent flexibility in both hardware and software aspects and can scaffold various learning activities.

1.5.1 LEGO Mindstorm Education NXT Base Set Robots

LEGO, leveraging off its vast experiences with building blocks for children's toys, collaborated with the Massachusetts Institute of Technology (MIT) to release the Mindstorm Education NXT Base Set robots package in September 2006. The package uses the building blocks concept to break down the hardware structure and at the same time incorporates sensing, motion, and control components to equip the robots with a high degree of flexibility and allow structural designs to rapidly construct intelligent robot prototypes. For those who do not have knowledge of the hardware layer, LEGO robots serve as an ideal educational tool. This package is even accessible by children, who can build their own hardware architectures. For these reasons, it has been used in this book.

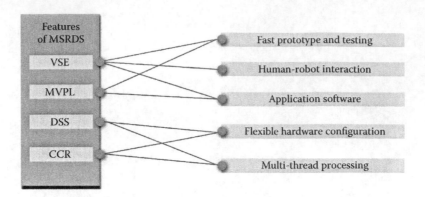

FIGURE 1.4   MSRDS satisfies the five main requirements of a service robot.

1.5.2 Microsoft Robotics Development Studio

Microsoft Robotics Development Studio (MSRDS) is a product that Microsoft introduced to the field of robotics in 2006. The product was first named Microsoft Robotics Studio, or MSRS, and later changed to MSRDS in 2008. It has the following features:

1. It supports Concurrency and Coordination Runtime (CCR) and Decentralized Software Services (DSS). This allows users to easily and flexibly control concurrent operations.

2. It supports the Microsoft Visual Programming Language (MVPL) environment. This provides a high-level graphical interface that is very accessible to those who have limited software engineering experience, allowing users to integrate various software modules easily.

3. It supports the Visual Simulation Environment (VSE). This allows designers to test and develop algorithms at times when no robots are available for testing.

The above features of MSRDS satisfy the five main requirements of a service robot: flexible hardware configuration, multi-thread processing, fast prototyping and testing, human-robot interactions, and application software. The relationships between these requirements are shown in Figure 1.4. Therefore this book uses MSRDS as the platform for teaching intelligent robot software development. The next chapter begins with a full introduction of MSRDS. The four major features of MSRDS—that is, CCR, DSS, MVPL, and VSE—are explained in more detail in Chapter 2.

1.6 BOOK STRUCTURE

Chapters 2 to 4 introduce the development environment of MSRDS. Chapter 2 provides a general description and explains the installation procedures. Chapter 3 introduces the Microsoft Visual Programming Language (MVPL), and Chapter 4 introduces robot simulation. In Chapters 5 and 6, we introduce the inputs and outputs to the robot and the control logic. Chapters 7 and 8 describe how MSRDS can be used to control a LEGO robot's hearing and vision. Chapter 9 describes a real-life teaching case example—the Robot Sumo contest. The final chapter provides other relevant information, as an extension of this book, for interested users.

CHAPTER 2

Microsoft Robotics Development Platform

2.1 ORIGINS OF DEVELOPMENT

In the January 2007 edition of *Scientific American Magazine*, the founder and chairman of Microsoft Corporation, Bill Gates, emphasized that "[t]he robotics industry will become a hot topic in the next three decades."* The current stage of robot development resembles the computer industry of the 1970s, as today's commonly used large industrial robots (such as computer-equipped industrials or military mine-detecting robots) are equivalent to the large-scale mainframe computers of those years. Eventually, computer development progressed toward miniaturization and became increasingly portable and lifelike. It is predicted that the robotics industry's next robot revolution will set off waves of service robots. In recent years, products such as robotic toy dogs, LEGO robots, and floor-sweeping robots have entered consumer markets. This is evidence of the early stages of robots becoming more integrated into our lives, not merely as a form of family entertainment, but with possibilities for bringing about significant changes to the way we live.

Although the number of robotics research teams has been growing rapidly with fruitful results being achieved, the current generation of robots lacks standardized hardware and software platforms—just as the large-scale computers did back in the 1970s. As the specification varies for each type of robot and their installed programs cannot be shared, software components often have to be written from scratch for the development of new types of robots. It also follows that the knowledge built in the development of a robot cannot be passed on. To reverse this situation, Microsoft Corporation began research and development on a robot software platform, and in 2006 the Microsoft Robotics Developer Studio

---

* W. H. Gates said: "The emergence of the robotics industry, which is developing in much the same way that the computer business did 30 years ago ... as these devices become affordable to consumers, they [the robot] could have just as profound an impact on the way we work, communicate, learn and entertain ourselves as the PC has had over the past 30 years." In "A Robot in Every Home," Scientific American Magazine, edited by Scientific American, January, 2007.

(MSRDS) was released. Apart from alleviating the above problems, its aim was to provide a common platform and basic service components to allow every designer with an interest in developing robots to easily program the necessary procedures to run robot hardware, without being restricted by the underlying complexity of the hardware system.

To address the concurrent-processing problem of driving both a robot's sensors and motion components, and to simplify the programming of robots and promote software reuse, the MSRDS development team incorporated various new programming concepts and functionalities into the software design. This allows MSRDS to control not only the robot but also any other service-based equipment. Furthermore, MSRDS is tightly coupled with the Microsoft Windows® operating system. As Microsoft Windows supports a wide range of software, MSRDS can easily be incorporated into the code base of Windows, making development easier and more flexible. Apart from this, the four design features—namely Concurrency and Coordination Runtime (CCR), Decentralized Software Services (DSS), Visual Simulation Environment (VSE), and Microsoft Visual Programming Language (MVPL)—facilitate robotic developments using MSRDS. In the remaining sections of this chapter, we describe CCR, DSS, and VSE. We explain MVPL and its application in teaching in Chapter 3.

2.2 CONCURRENCY AND COORDINATION RUNTIME

Service robots are often faced with complex and changing environments. In such environments, service robots have to continuously collect information from the surroundings using sensors and to control actuators to respond accordingly to situations. This coordination of tasks requires concurrent programming technology and cannot be handled through a sequential programming model.

Let's take the classic example in which a robot constantly surveys its surroundings while walking to prevent itself from falling. The typical approach is to use multiple sensors to collect environmental information (as shown in Figure 2.1). Here, a sequential program would wait until it has collected information from all the sensors, construct a model of the environment, and then initiate an action on the actuators accordingly. However, if one of the sensors has an overly slow response or becomes faulty, the sequential program would

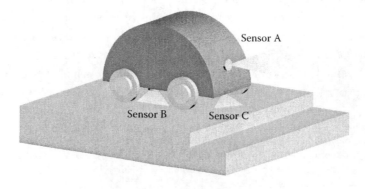

FIGURE 2.1   A robot requires multiple sensors to detect if it will fall.

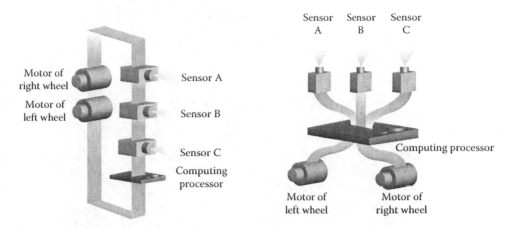

FIGURE 2.2   The process differences between a traditional sequential program (left) and a concurrent program (right).

halt or be adversely affected with the result being a substantially increased risk of the robot falling. If the same situation is handled by a concurrent processing program, the system can receive information simultaneously from multiple sensors and the entire system can continue operating even in the case of a failed sensor as the information gathered from the other sensors can still be used to generate a response for the actuator (as shown in Figure 2.2).

The concurrent programming technique is not uncommon, and the multi-threaded processing technique used in operating systems of personal computers is a ubiquitous example of this. However, one difficulty is that the technical skill level required for this kind of implementation is prohibitively high, to the point that even experts who have a computer science and engineering background may find difficulty in applying it to construct a robot control system. One of the beneficial features of MSRDS is that it provides concurrent programming through CCR, enabling programmers to write concurrent programs easily and intuitively as well as making debugging and data management more convenient. Furthermore, with the development and popularization of ever-smaller computing units, most computer chips have integrated sensing and processing components to reduce the processing overhead of the sensing components on the system. For these designs, concurrent processing can be applied to create a more efficient robot control system.

2.3 DECENTRALIZED SOFTWARE SERVICES

DSS is a lightweight service-oriented model, used to address event scheduling and the asynchronous processing problem between sensors. In MSRDS, DSS utilizes the *state* orientation mechanism to coordinate asynchronous processes, as shown in Figure 2.3. *State*, as its name suggests, refers to each parameter of the robot service component, examples of which include the map information, sensory values, and a robot's movement direction. The state in the MSRDS platform is unique. When any service component changes the state, other service components will observe the latest state information and need not worry that the system information is not up-to-date.

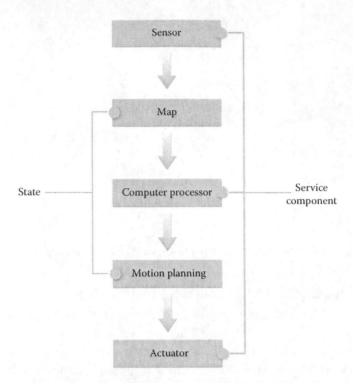

FIGURE 2.3 DSS service-oriented model can easily coordinate service components and the robot's state.

As an example, *map* and *motion mode* are important items of information in the robot's system and they can be recorded in the state. At the same time, the system may have many service components, including a path-planning component (a component packaged with the path-planning algorithm) that is responsible for optimal path calculation, an actuator component responsible for controlling motion, an acoustic sensing component responsible for distance calculation, and an imaging component responsible for recognizing environmental images. DSS allows each service component to update information such as the map and the motion mode in the state at any time without interference. For instance, the imaging component can collect information about changes in the environment, while the acoustic component confirms the distance between surrounding objects, and simultaneously, the two components can work together to continually update the map. Furthermore, the path-planning component can use the map at any time to calculate the most suitable motion mode and the actuator can use the motion mode at any time to drive the robot's motion (Figure 2.3). This is the concept and convenience that DSS brings.

In addition, the DSS service components can be packaged and ported to another robot's control program for reuse. What's more, in the design of complex robotic systems, programmers no longer have to face many different hardware designs and no longer have to agonize over how to assign several simultaneous tasks. This is the highly object-oriented programming design concept. DSS service components also include a network component. As a result, MSRDS programmers can easily use XML to access a robot's state information and publish it online for sharing with other programmers (Figure 2.4).

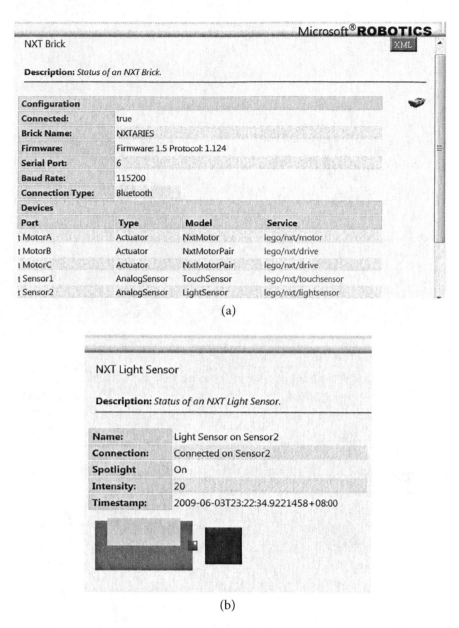

FIGURE 2.4 A browser can access and display the robot's state information under the DSS framework: (a) The list of DSS services; (b) the real-time status of the light sensor.

2.4 VISUAL SIMULATION ENVIRONMENT

At the research and development stage of a robot, the acquisition cost of robot hardware components cannot be ignored. If a virtual environment can be used to simulate a physical robot's response, we can develop algorithms before purchasing the robot to test the robot's limitations. This in effect allows robotics education to be achieved at a much-reduced cost and enhances students' interest. For this purpose, Microsoft utilized its rich experience in the gaming industry from the game development platform XNA 2.0 (running on

FIGURE 2.5   A screenshot of the VSE provided by MSRDS.

Microsoft's Xbox 360) to develop the Visual Simulation Environment (VSE). VSE allows MSRDS developers to test the interactions between several robots and obstacles in a virtual simulation environment. As XNA 2.0 is specifically designed for graphic-intensive games, the simulation environment in MSRDS is much more graphically rich than the environment of other robot systems.

To enhance VSE's graphical performance and make the actions of the simulated robots seem more realistic, Microsoft has utilized the physics engine developed by AGEIA to provide physics for the simulator. Apart from being able to use the CPU to make calculations, the physics engine can also utilize the physics graphics card developed by the company as a supplement. In 2008, AGEIA was acquired by the well-known graphics card manufacturer NVIDIA, and hence any NVIDIA physics graphics card can be used to bolster the performance of the simulation in VSE (Figure 2.5).

2.5 MICROSOFT VISUAL PROGRAMMING LANGUAGE

For the development stage, MSRDS provides two choices: (1) use of the Microsoft Visual Studio with the C# programming language; or (2) use of the Microsoft Visual Programming Language (MVPL). Since the first choice (which is also the primary development environment for MSRDS) is more difficult for application designers of service robots, the second choice (i.e., MVPL) is recommended in this book. Its interface incorporates an unconventional programming method, in that it does not emphasize written code, but rather a high-level graphical user interface to complete MSRDS programming. Typical programmers can utilize MVPL to speed up the pace of learning, whereas advanced programmers can utilize MVPL to rapidly test their algorithms and display the logic (Figure 2.6). As MVPL

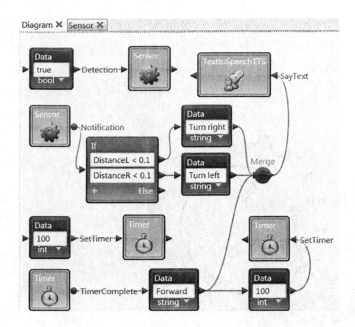

FIGURE 2.6 MVPL uses an intuitive and high-level graphical interface for programming.

involves details of actual programming, a detailed explanation is provided in Chapter 3 of this book.

As described above, MSRDS provides four design features: Concurrency and Coordination Runtime (CCR), Decentralized Software Services (DSS), Visual Simulation Environment (VSE), and Microsoft Visual Programming Language (MVPL). These features effectively lower the knowledge requirements for application experts to enter the domain of robotic design. Furthermore, MSRDS provides an integrated development platform that is beneficial for work collaboration between robotics development staff.

2.6 SYSTEM REQUIREMENTS

MSRDS can be installed on all versions of the Microsoft Windows operating system, including Windows CE,* Windows XP, Windows Server 2003,† Windows Vista, and Windows 7. However, readers are advised to install it on either Windows Vista or Windows 7 to utilize human-computer interaction features, such as speech recognition, that are built into the operating system. Visual Studio 2005‡ and above can be used as the editing environment. In addition, if the VSE is to be used, Microsoft's multimedia library DirectX 9.0 or above must be installed. The hardware requirements are reasonably achievable in currently available mainstream desktops and notebooks.

---

* The installation of MSRDS on Windows CE only limits use to MSRDS developed applications and development on the platform is not supported.
† Microsoft Windows Server 2003 must have Service Pack 2 or above for MSRDS installation.
‡ Microsoft Visual Studio 2008 is advised.

2.7 INSTALLATION

This book employs the **MSRDS 2008 R3** Edition, which is freely available **from** the official website at http://www.microsoft.com/robotics/. Click on **the** link at "**NEW—Get Microsoft Robotics Developer Studio 2008 R3**" at the top of the page and **Get It Now** on the next screen to enter the **Download Center** page.

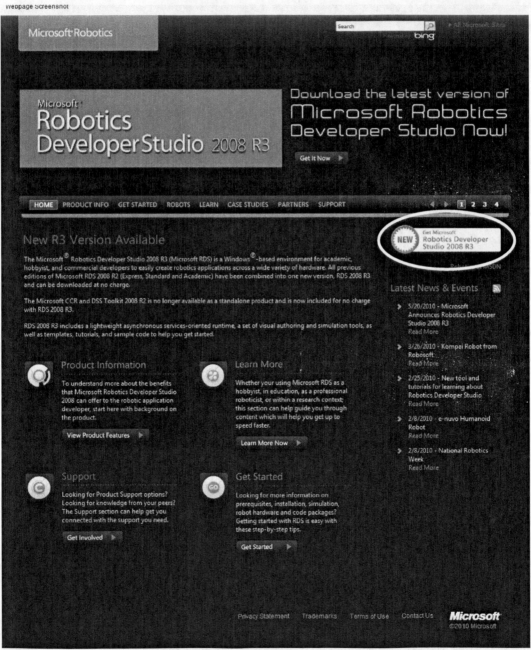

http://www.microsoft.com/robotics/

Webpage Screenshot

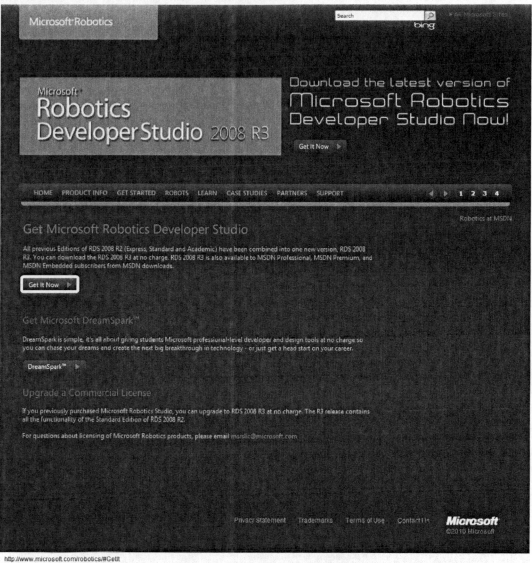

After entering the Downloads page, you can then click on **Download** to get the install package for MSRDS 2008 R3.

After the download is complete, proceed with the installation. The program will display the installation screen. Following this, please click on **Next**.

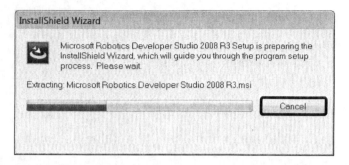

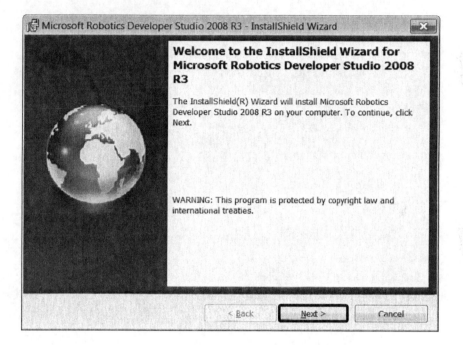

Next, the MSRDS license agreement is displayed. Please check the box next to "**I accept the terms of the license agreement**".

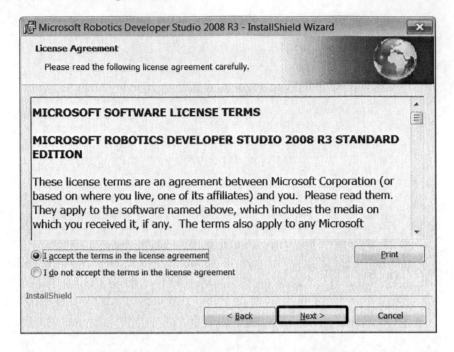

The MSRDS Microsoft **Robotics Feedback and Update Programs** notice is displayed. You can join the feedback program and enable automatic update checks.

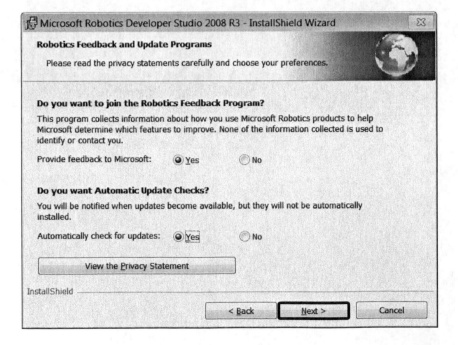

In the next step, choose the type of setup. Here, we choose the **Complete** setup.

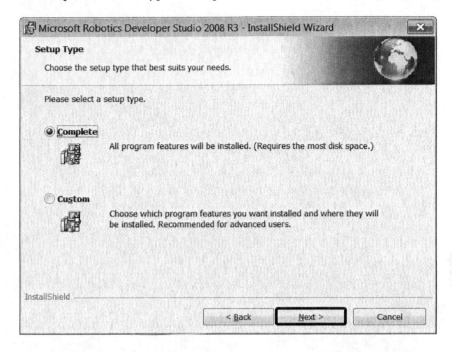

Following this, the confirmation page is displayed. Click **Install** to begin the installation.

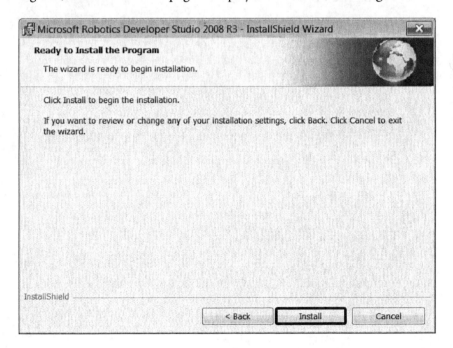

Wait for the installation to complete. It usually takes between two to five minutes, depending on the hardware configuration.

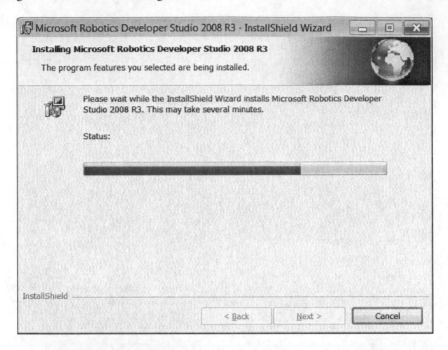

After the installation is complete, click **Finish** to end the MSRDS installation.

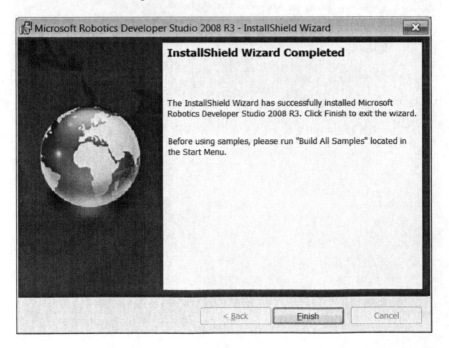

At this point, you can locate the software in the **Start Menu** and can run the Microsoft Visual Programming Language (MVPL). In the following chapter, we introduce MVPL.

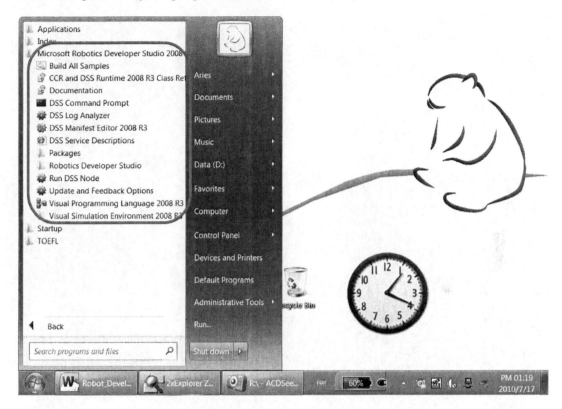

This chapter provided the instructions for installing MSRDS. Before proceeding, you should confirm that the various components of MSRDS have been installed and that the MVPL application can be opened. The programs installed will be used in the teaching examples in the following chapter, which are used to describe how to write simple robot programs.

2.8 MSRDS PROGRAM LIST

After the installation of MSRDS 2008 R3 Edition is complete, several procedures are produced. Each procedure's function is described as follows:

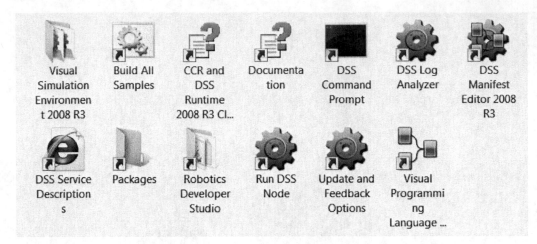

This image has been reproduced exactly as captured on screen.

Visual Simulation Environment 2008 R3

VSE's examples folder contains several VSE example files. The user can open each example to understand the VSE runtime environment. All the examples are usable in robot simulations.

Build All Samples

This is a separate manual step which allows you to build all samples attached with MSRDS.

CCR and DSS Runtime 2008 R3 Class Reference

CCR and DSS documentation allows the user to query the classes and methods within CCR and DSS. The documentation also contains detailed usage descriptions, which greatly aid the program designer in the development of service components.

Documentation

MSRDS documentation includes an MSRDS introduction, system architecture, and teaching files (including CCR, DSS, MVPL, and VSE) in addition to others. These are essential reference files for using MSRDS and software development.

DSS Command Prompt

DSS's command-line prompt is used to assist the user in operating DSS in the command mode environment

DSS Log Analyzer

You can use the DSS Log Analyzer tool to view log files recorded using DSS logging, such as messages and filter message flows across multiple DSS services. You can also inspect message details.

DSS Manifest Editor 2008 R3

The user can use the manifest editor to edit the configuration of hardware components. Further related information is discussed in Chapter 6 of this book.

DSS Service Descriptions

An introduction document to the DSS service description directory, which provides a simple mechanism for announcing DSS service descriptions using RSS technology and is compatible with common RSS readers.

Packages

Packages is a default folder for other plug-in components, which are robot service components that users can obtain from the internet or from third-party sources. The components are installed in this folder for use.

Robotics Developer Studio

The MSRDS installation folder contains all MSRDS files (documentation, service component file, each example's source, and so forth).

Run DSS Node

DSS service platform is used to start the most primitive DSS service. After starting, all MSRDS service components can be started or stopped via the dynamic webpage (The default address is http://localhost:50000/).

Update and Feedback Options

This item collects information about the use of MSRDS, in order to help Microsoft improve the program.

Visual Programming Language 2008 R3

MVPL's editor provides the user with an easy-to-learn and intuitive software development environment. Detailed explanations are provided in Chapter 3.

CHAPTER 3

Microsoft Visual Programming Language

3.1 OVERVIEW

The purpose of this chapter is to describe the style of programming under Microsoft Visual Programming Language (MVPL) and, through examples, describe how to write sequential and concurrent programs. All examples are described with a combination of text and diagrams to guide the reader into understanding MVPL's programming environment.

3.2 MVPL PROGRAMMING ENVIRONMENT

Traditional programming interfaces are mainly text-based (Figure 3.1a), and as such, they pose a degree of difficulty when used as the program development environment for complex robots. The visual programming language (VPL) concept has existed for a long time and can help the programmer to reduce this difficulty. It emphasizes the use of graphics, tables, or a combination of the two for interface control to lower the programming difficulty and allow programmers to focus more on the logic, rather than the written code itself. Several programming environments have already embraced the VPL concept, such as National Instrument's engineering and scientific programming software LabVIEW, MathWork's Simulink, Apple's Automator and Quartz Composer, and the art and design software Scratch, among others. MSRDS also uses the VPL concept, which is called the Microsoft Visual Programming Language (MVPL) interface, to reduce some of this difficulty (Figure 3.1b).

MVPL allows MSRDS, in the design of robot programs, to apply the concept of drag-and-drop design components for generating and building connections. The design component refers to either basic data or a more complicated service. MVPL's environment can be divided into the command, basic activity component, service component, program diagram, project, and component properties panels (Figure 3.2). Diagrams are used to display

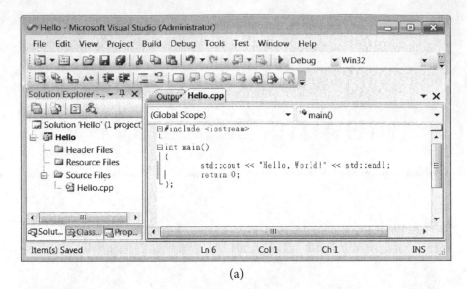

(a)

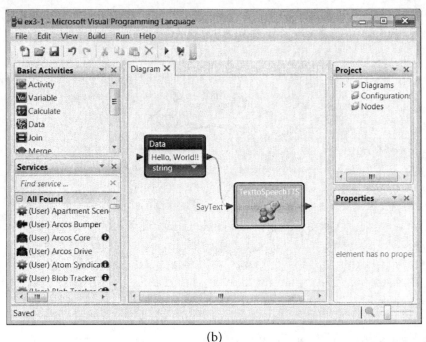

(b)

FIGURE 3.1 The differences between text-based and visual programming interfaces: (a) a text-based programming interface (using Microsoft Visual Studio editor as the example); (b) a visual programming interface (using the Microsoft Visual Programming Language 2008 R3 Edition as an example).

the program logic, and the user can drag-and-drop an activity component or a service component into the diagram area. This enables the user to create a component (Figure 3.3) as well as to build the connections intuitively with the data flow or direction of execution, whereby different properties of components can be combined as a means to express the logic behind a program's operations and complete its design (Figure 3.4). During program

basic activity component panel · · · · command panel · · · · project panel

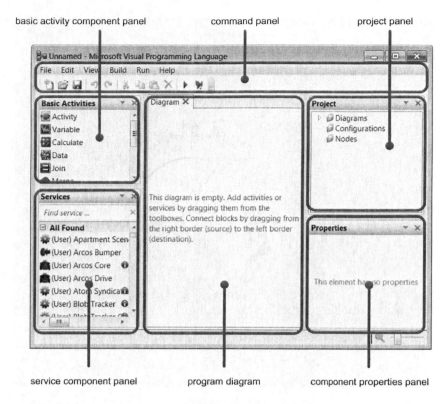

service component panel · · · program diagram · · · component properties panel

FIGURE 3.2 A screenshot of the MVPL programming environment. The blue (i) icons beside the services in the services panel provide additional information by linking to the Microsoft Developer Network (MSDN) website (http://msdn.microsoft.com/).

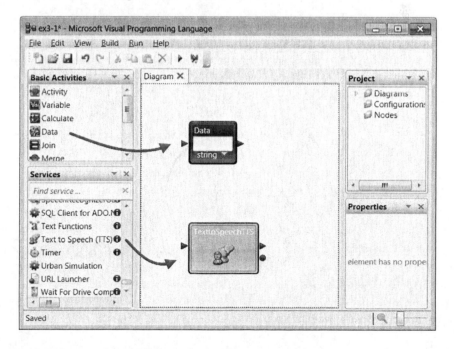

FIGURE 3.3 MVPL uses drag-and-drop to create components.

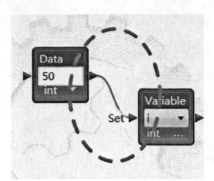

FIGURE 3.4   MVPL uses lines to build connections between components, representing the data flow or the direction of execution.

compilation, MVPL will appropriately remind the user of the necessary setup parameters or procedures before building components or connections. Furthermore, the user can also modify specified properties after the components and connections have been built.

Upon the completion of program compilation, the user can select **Start** under **Run** on the command panel or press the <F5> key to execute the program. If you wish to monitor each step of the program execution flow, you can select **Debug** under **Run** on the command panel or press the <F10> key. The program will then enter the debug mode through the web browser interface and the variable values at each program step can be observed (Figure 3.5).

3.3 BASIC ACTIVITIES AND SERVICES COMPONENT

Every component within the basic activities component represents a variable or the most basic logic concept, while the service component represents the connections with MSRDS external programs. As shown in Figure 3.6, the left side of each component defaults as the data input end while the right side defaults as the data output end, marked by the red triangles correspondingly. In the figure, the direction in which the triangles are pointing represent the direction of data transfer (input to or output from the component). Users can use the markings on the red triangles to drag-and-drop line connections to build connections between components. Apart from the red triangle at the output end, a red circle may also be displayed in some components. The output of those components can thus be divided into the Output and the Notification end. The Output end outputs the calculated result by the component after processing one or more input values (the input point arrow as the point of data flow), while the Notification end serves as a data output point to notify that an event has happened, transmitting the default notification message when there is component activity. The Notification end is represented by a circle and is able to notify other components when the component information has changed. It is worth noting that the Notification end occurs asynchronously but the Output end only arises in response to an input message.

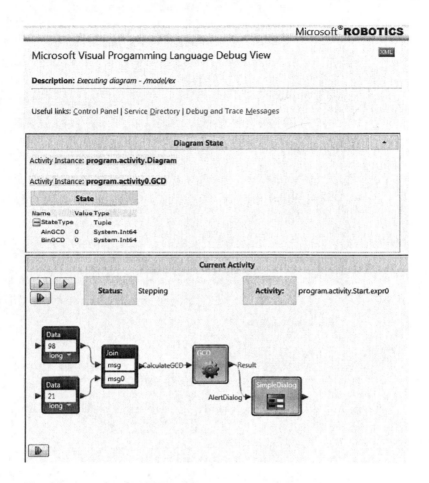

FIGURE 3.5   The debug mode of MSRDS.

FIGURE 3.6   A component's data input and output ends.

The significance of every basic activity component is described as follows:

Variable—Variable components are used to store values for use by other activity components. There are two functions related to the Variable component: (1) SetValue, which is used to set the variable value within the Variable component; and (2) GetValue, which is used to retrieve the variable value for use by other activity or service components.

Calculate—Calculate components are used to perform simple arithmetic calculations and logic. The logical operators include the && symbol representing AND, the || symbol representing OR, and the ! symbol representing NOT. Apart from these, Calculate can be used to access individual messages in input data, for example, retrieving individual button presses on the Xbox 360 controller.

Data—Data components are used to set the initial value of Variable or the input value to the Calculate component. The type can be any datatype supported by .Net.

List—List components are used to establish a list of some particular datatype to be used by other activity components.

List Functions—List Functions components are used to perform operations on items stored in the List component. These operations include Append, Concatenate, Reverse, Sort, RemoveItem, InsertItem, and GetIndex.

Join—Join components are used to join multiple input values into one output value. This component will only output a message when both input values are present.

Merge—The purpose of the Merge component is to funnel multiple message sources into a single input. It is often confused with the Join component by novice users. Although Merge components support multiple input values and only produces one output, an output can be produced even with one input value. This is the main difference between the Merge and the Join component.

If—Decision components have the same meaning as the traditional program construct "If … then … else …." Its syntax includes "equal" (represented by two equal signs, ==), "less than" (represented by <), "larger than" (represented by >), and "not equal" (represented by !=).

Switch—Switch components have their output determined by the input value and the decision module. Their function is similar to the Switch syntax in the C++ programming language.

Comment—Comment components are used to add comments to activity components.

Custom Activity—The Custom Activity component is used to establish user customized activities. This component allows the user to set the contents of the activity. Other activities and services can be used within a newly established activity. Every user's customized activity can receive input and return output, as well as trigger notification events.

3.4 BASIC PROCESS CONTROL IN A PROGRAM

In the previous section, the functionalities of the basic activity components were described. In this section, we use several examples to explain how simple application programs can be built.

EXAMPLE 3.1: HELLO WORLD VOICE COMPOSITION

Explanation: Enables the computer to read out "Hello, World."

Skill: The skill required in this example is the ability to construct a basic activity component and a service component, as well as knowing how to construct the connection between the two.

Completed diagram:

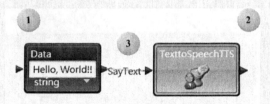

(The numbered balls represent the steps described below.)

TIP

"Hello World" is the classic introductory task in many programming lessons as it concerns the concepts of basic input, output, and variable value storage. For a typical program, the Hello World example normally involves printing out "Hello, World!!" on screen. However, given MSRDS's support for voice construction, this Hello World example uses voices to produce the "Hello, World!!" sound.

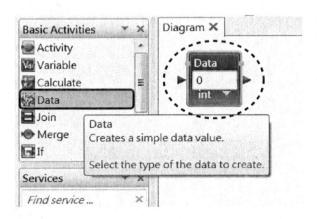

Step 1a—Drag-and-drop the **Data** component template into the **Diagram** area to create a new **Data** component. (Double-clicking on the **Data** template component using the left mouse button achieves the same effect.)

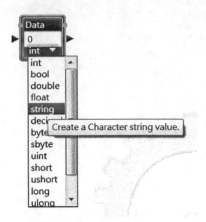

Step 1b—Click on the drop-down menu arrow and change the data-type to **string**.

Step 1c—Type in "Hello, World!!" in the **Data** component and this sets the **string** to "Hello, World!!"

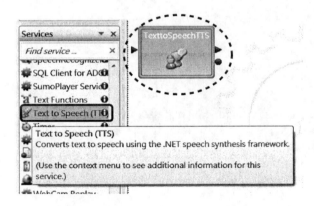

Step 2—Construct a voice formation service component within MSRDS known as **Text to Speech** by drag-and-dropping from the **Services** area.

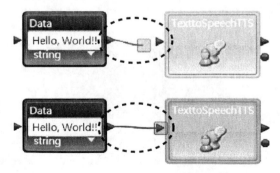

Step 3a—Use the drag-and-drop functionality to connect the triangle on the **Data** component to the input end of the **TextToSpeechTTS** component, establishing their connection.

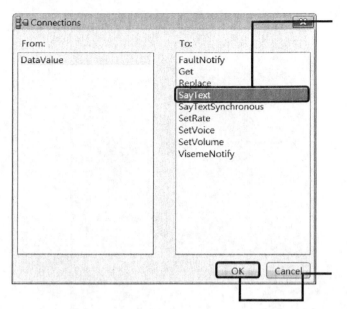

Step 3b—While building the connection, the **Connections** dialog box will pop up. Select the **SayText** function to designate this connection as reading out information using voice.

Step 3c—Click on **OK** to store this setting.

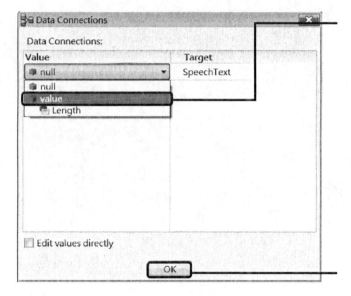

Step 3d—In the next pop-up box, change the data type to **value** to set the input value as the data itself. The other menu option here is **Length**, which describes the data length.

Step 3e—Click **OK** to store this setting.

After completing the above steps, the example is complete. Press the <F5> key to start the DSS service (Figure 3.7). The software will then begin compilation and execution of the program, speaking the two words "Hello" and "World," as the **Text to Speech** component can only create text-based speech.

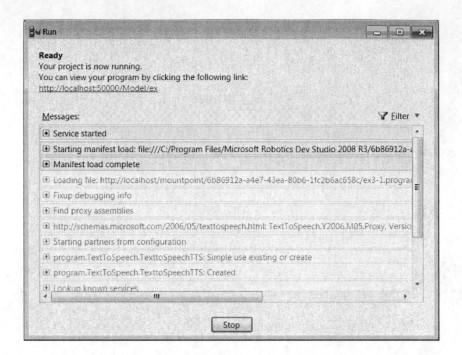

FIGURE 3.7   MSRDS will invoke the DSS service during execution.

EXAMPLE 3.2: INCREMENT

Explanation: Enables the computer to increment its counter ten times, saying out the value after each increment. At completion, the computer will say, "Done."

Skill: The user is required to create a repeated loop and combine the **Text to Speech** component from example 3.1.

Completed diagram:

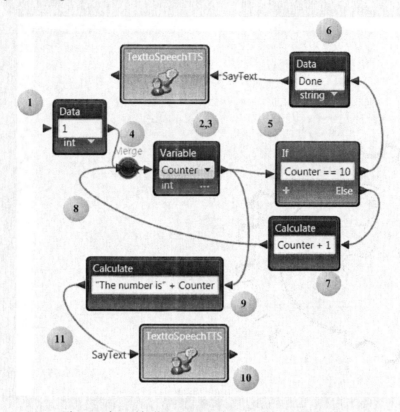

(The numbered balls represent the steps described below.)

 Step 1—Create the basic component **Data** as the initial value of the whole program. Set the type as **int** and its value as **1**.

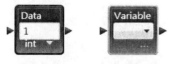 Step 2—Create the basic component **Variable**, used to store the variable.

 Step 3a—Click the lower right corner of the **Variable** component to set variable characteristics.

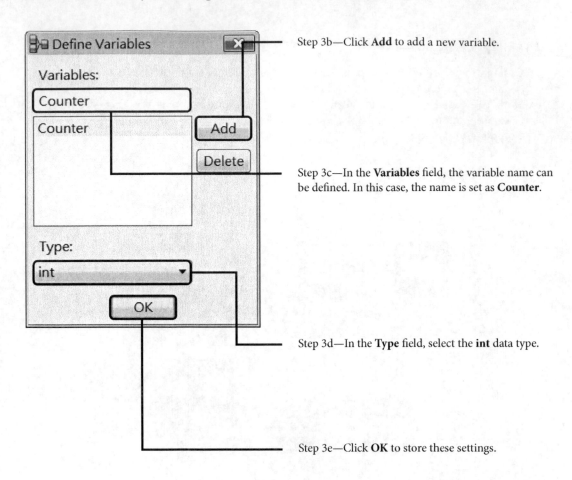

Step 3b—Click **Add** to add a new variable.

Step 3c—In the **Variables** field, the variable name can be defined. In this case, the name is set as **Counter**.

Step 3d—In the **Type** field, select the **int** data type.

Step 3e—Click **OK** to store these settings.

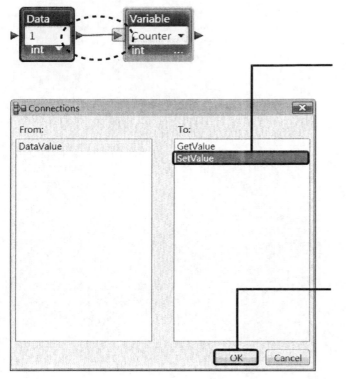

Step 4a—Drag a line between the **Data** component and the **Variable** component to build the connection between them. In the connections setting window that pops up, choose **SetValue** to assign the value of **Data** to **Variable**.

Step 4b—Click **OK** to store this setting.

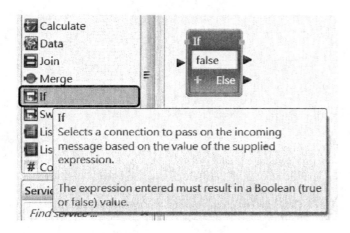

Step 5a—Create an **If** basic activity component.

Step 5b—Create the connection between **Variable** and **If** components and enter **Counter == 10** into the condition field.

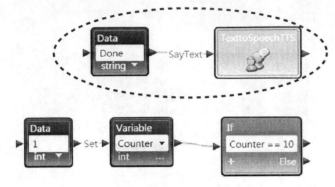

Step 6a—Create **Data** and **Text to Speech** components as the program completion actions, so that at program completion it reads outs "**Done.**"

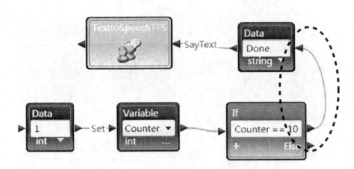

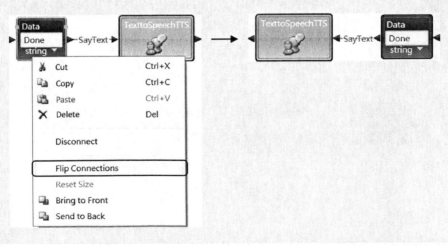

Step 6b—Create a connection between the right output end triangle of the **If** component and the action terminating **Data** component. This means that once the **Counter == 10** condition is satisfied, the word "Done" will be read out.

TIP

To see the program flow chart more clearly, right click on basic activity or service components to reveal the **Flip Connections** menu option. This option flips the input and output ends, adjusting the component presentation without any changes to its functionality. The example below shows the **Data** and **Text to Speech** components following this change.

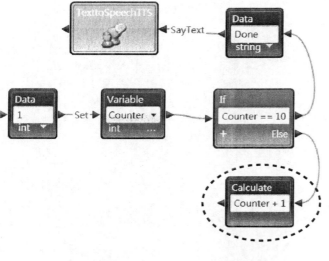

Step 7a—Create a **Calculate** component and create a connection with the lower right triangle of the **If** component. This means that when the **Counter** does not equal 10, the data will be outputted to the **Calculate** component.

Step 7b—In **Calculate**, enter the text **Counter+1** (spaces are not required around the +). This means that if Counter is not equal to 10, the value of **Counter+1** is calculated. This does not change the original variable value of **Counter**.

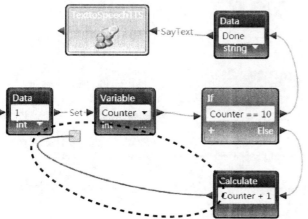

Step 8a—To update the value of **Counter**, the connection between the **Calculate** and **Variable** components needs to be created. This requires a line between the output end of **Calculate** and the input end of the **Variable** component that stores the **Counter** variable. As **Variable** has been initialized by a **Data** component, the connecting line should be drawn in addition to the **Set** line between **Data** and **Variable** components.

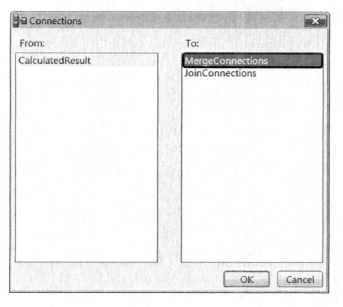

Step 8b—In the **Connections** dialog box that pops up, select **MergeConnections**.

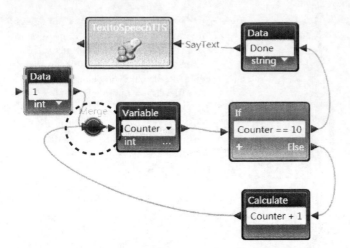

Step 8c—A new **Merge** object is automatically created. As the **Calculate** component calculates **Counter+1**, it can be used to set the new value for the **Counter** variable. This is the loop creation method in MVPL.

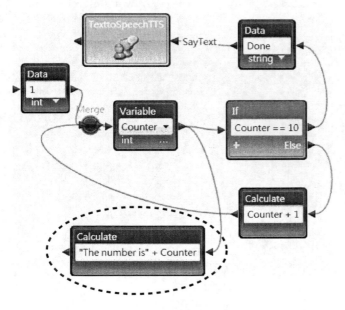

Step 9—To allow the **Variable** component to report **Counter**'s new value at every increment, create a new **Calculate** component and connect the output end of the **Variable** component to direct output to the input end of the **Calculate** component at the same time. Type in **"The number is"** + **Counter** for the calculation formula of the **Calculate** component.

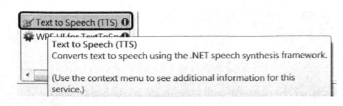

Step 10a—Add a new **Text to Speech** component.

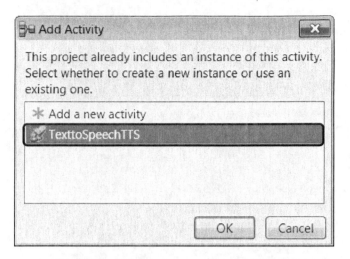

Step 10b—As this is the second **Text to Speech** component, a dialog box will appear allowing the user to confirm whether the existing **Text to Speech** component will be used or that a new component needs to be created. Although neither selection would affect the execution outcome in this example, here we choose to use the existing **Text to Speech** component to ensure that they have the same state in the procedure of the whole program.

TIP

When writing programs in the MVPL environment, if several service components with the same type are to be created (for instance, given an existing **Text to Speech** component, other **Text to Speech** components are to be created), a window will pop up asking if a new service component is to be created or if an existing component is to be used (where there can be two service components with the same name). This is to allow the programmer to conveniently write the code for the DSS concept and simplify the program logic.

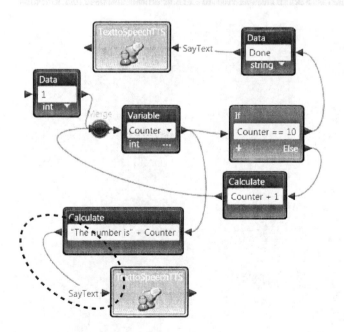

Step 11—Create the connection between the **Calculate** component "**The number is**" + **Counter** and the **TexttoSpeechTTS** component so that the **Counter** value is reported at each update. For instance, when the **Counter** has been updated to 7, the program would report "The number is 7." At this point, the program is complete.

After completing the above steps, the example is complete. Press the <F5> key on to start the DSS service. The execution result of the compiled program should report in sequence "The number is 1," "The number is 2," and so on; when "The number is 10" is reached, the word "Done" will be spoken.

EXAMPLE 3.3: OBTAINING THE GREATEST COMMON DIVISOR

Explanation: Using the method of successive division (also known as the Euclidean algorithm), this example obtains the greatest common divisor of two numbers. The steps to obtain the greatest common divisor between a and b are as follows:

1. Calculate $a_1 = a / b$
2. Calculate $b_1 = b / a_1$
3. Calculate $a_2 = a_1 / b_1$
4. Repeat the division until we obtain $a_n = b_n = c$, with the greatest common divisor being c.

Skill: The skill required in this example is the ability to create a custom activity component.

Completed diagram:

1. Overall program

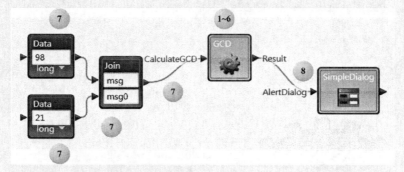

2. Custom activity component—greatest common divisor

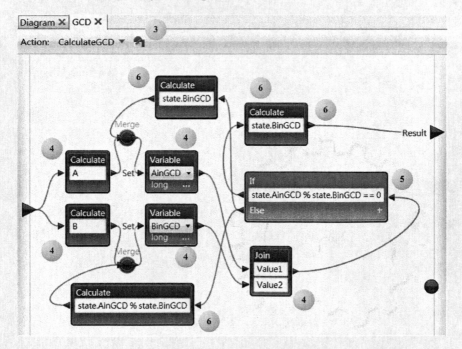

(The numbered balls represent the steps described below.)

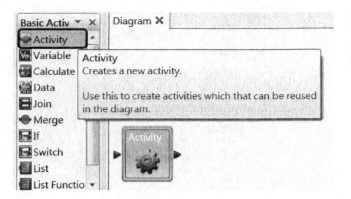

Step 1—Create an **Activity** component to package components used for calculating the greatest common divisor.

Step 2a—After selecting **Activity**, select **Properties** at the right end of the main program window in MVPL to modify its properties. Modify its name (**Name** field) to **GCD**. This is the abbreviated form of the greatest common divisor. The full term **Greatest Common Divisor** is to be typed into the **Friendly Name** field. Customizing the names does not affect functionality and only serves the purpose of allowing users to distinguish among other created components.

Step 2b—**GCD** is shown, indicating that the name change is complete.

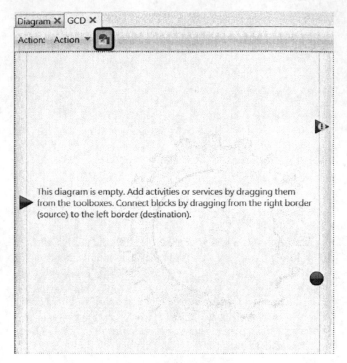

Step 3a—Double left-click on the **GCD** component to switch between internal editing screens.

Step 3b—Select the top left corner of the **GCD** component edit window as indicated to edit the input and output items.

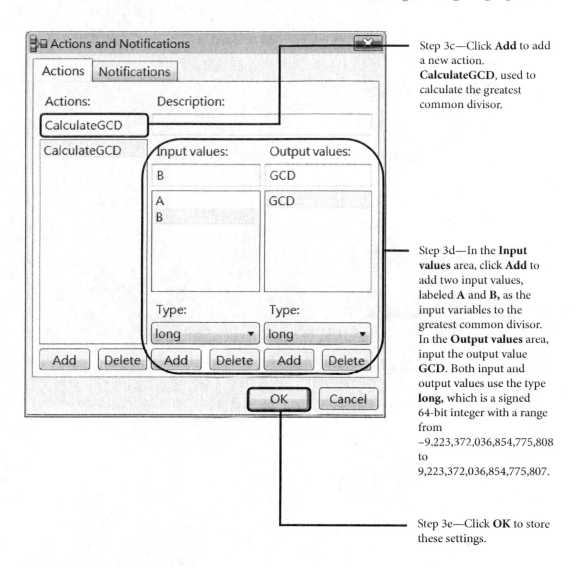

Step 3c—Click **Add** to add a new action. **CalculateGCD**, used to calculate the greatest common divisor.

Step 3d—In the **Input values** area, click **Add** to add two input values, labeled **A** and **B,** as the input variables to the greatest common divisor. In the **Output values** area, input the output value **GCD**. Both input and output values use the type **long,** which is a signed 64-bit integer with a range from −9,223,372,036,854,775,808 to 9,223,372,036,854,775,807.

Step 3e—Click **OK** to store these settings.

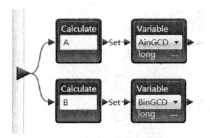

Step 4a—Return to the edit screen of the **GCD** component and create two variables **AinGCD** and **BinGCD,** both of the type **long**. The initial value will be provided by the two **Calculate** components. Connect the input ends of the **GCD** components **A** and **B**.

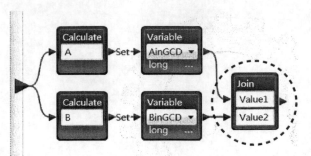

Step 4b—Use the **Join** component to combine the two **Variable** components as a set of numerical values.

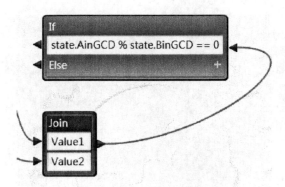

Step 5—Connect the **Join** component to the **If** component and use the condition **state.AinGCD % state.BinGCD == 0** to confirm whether the greatest common divisor can be determined. Here, the operator "%" indicates the division relationship, which in the component means the condition "AinGCD / BinGCD == 0."

TIP

The **Join** component is required to combine the variables **AinGCD** and **BinGCD** because an **If** component is unable to receive multiple input values. It is thus a rendezvous point for program control flow, as otherwise the messages from the two Sets would arrive asynchronously. Thus, **Value1** and **Value2** are not used here.

TIP

The term **state** here refers to the state information of DSS, which includes all of the variables in the program. It is conceptually similar to the namespace concept in object-oriented programming languages. While accessing the state internal information, a period (.) symbol needs to be added in between the state and the variable name.

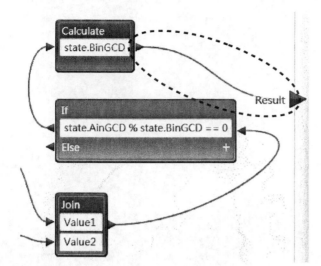

Step 6a—If the condition of the **If** component evaluates to true, it means that the greatest common divisor has been found. The **Calculate** component can then extract **state. BinGCD** from the **GCD** component is the greatest common divisor.

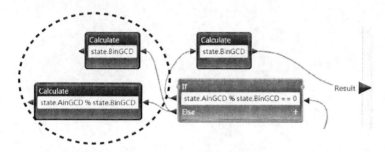

Step 6b—If the condition of the **If** component evaluates to false, it means that the greatest common divisor has not been found. The **Calculate** component then calculates and updates the values of **AinGCD** and **BinGCD**.

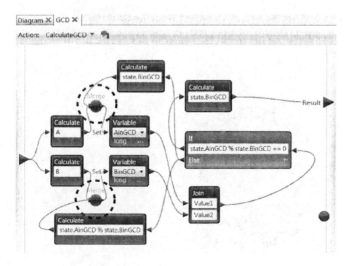

Step 6c—Using **Merge**, connect the resulting calculated value of the **Calculate** component to the two variable components **AinGCD** and **BinGCD**, completing the logic design of the **GCD** component.

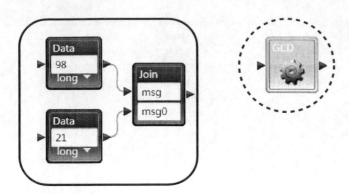

Step 7a—Returning to the main program, the **GCD** component is now complete. You can now use two values to test the **GCD** function. Here, input **98** and **21** into the **GCD** as a test. As **GCD** only accepts a single input, the **Join** component is used to combine two input values.

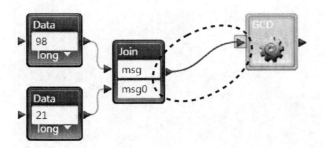

Step 7b—Create the connection between the **Join** component and **GCD** component. In the pop-up dialog box, we can select the input value relationship between the **Join** component and the **GCD** component.

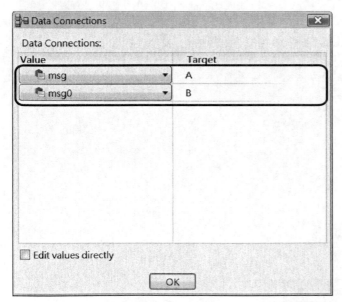

Step 8a—Use the **SimpleDialog** service component as the **GCD** output end connection component.

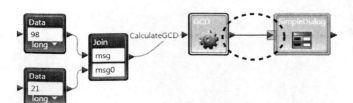

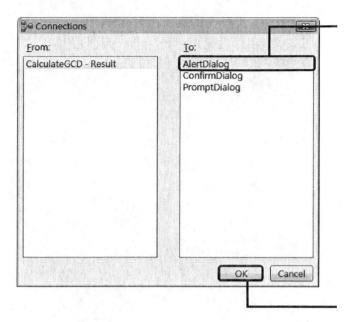

Step 8b—The **GCD** and **SimpleDialog** components connection state is set as **AlertDialog**. This results in the program using a pop-up window to output the result of the **GCD** calculation.

Step 8c—Click **OK** to store this setting.

TIP

The other two functions of the **SimpleDialog** service component are (1) ConfirmDialog, which, apart from displaying information, provides the two buttons **OK** and **Cancel** each with its own function; and (2) PromptDialog, which provides an input dialog box, allowing the user to enter required variable values.

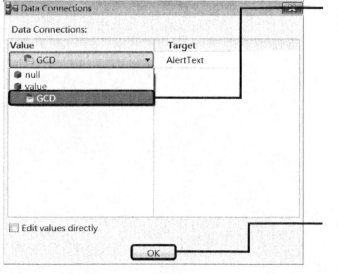

Step 8d—In **GCD**'s output value selection, select the **value** of **GCD**.

Step 8e—Click **OK** to store this setting.

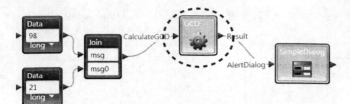

Step 9—As the completed program applies a customized component for **GCD**, it has been simplified because the complicated steps are encapsulated.

After completing the above steps, the example is complete. Press the <F5> key to activate the DSS service. The execution of the compiled program should result in a pop-up window showing the greatest common divisor as **7**.

TIP

For a typical text-based editing environment, simple calculation techniques are not difficult (like incrementing in example 3.2 or the greatest common divisor in example 3.3):

1. Incrementing program in C++

```cpp
#include <iostream>
using namespace std;
int main()
{
    for(int Counter=1; Counter<=10; Counter++)
        cout << "The number is " << Counter << endl;
    cout << "Done" << endl;
    return 0;
};
```

2. Greatest common divisor program in C++

```cpp
#include <iostream>
using namespace std;
long GCD(long AinGCD, long BinGCD)
{
    long temp;
    while(AinGCD % BinGCD != 0)
    {
        temp = BinGCD;
        BinGCD = AinGCD % BinGCD;
        AinGCD = temp;
    };
    return BinGCD;
};

int main()
{
    long A=98, B=21;
    cout << GCD(A, B) << endl;
    return 0;
};
```

Although MVPL appears to be more tedious and its execution possibly less efficient than other languages, its graphical interface is more intuitive and allows early adopters to learn quickly, while more advanced programmers can use it as a way to present the algorithm logic. Differing from sequential languages, MVPL adopts a concurrent programming environment. Implementing a sequential algorithm such as the greatest common divisor solver might be more difficult with MVPL than with a sequential language. However, MVPL is much more suited for designing service-type robot control systems, which need to coordinate various hardware and software. Apart from this, MSRDS has the advantage of inheriting the Microsoft Windows environment and provides a uniform interface (service component) for communicating with third-party hardware and software. This allows MSRDS to connect to them easily (for example, voice synthesis and the control of cameras connected to computers) while avoiding other functions that programs cannot action. Although C++ or the other programming languages can also do this, their implementation is much more difficult than MSRDS. This is also the reason why the MSRDS and MVPL environments are suited for the testing and development of service robot programs.

3.5 CONCURRENT PROGRAM CONTROL

MSRDS is based on the design of DSS, giving it concurrent programming capability. In example 3.3, we see that the data flow is scattered in execution and the scattered data flows are combined with a single **Join** or **Merge** component (Figure 3.8). This section uses two examples to demonstrate how concurrent programs can be implemented by using the MVPL interface.

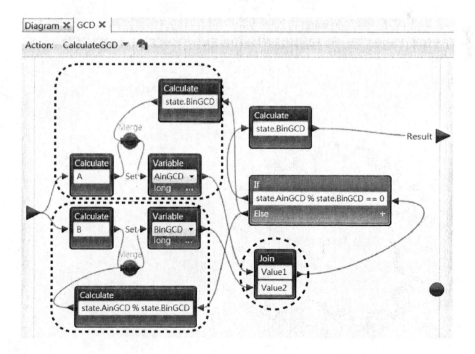

FIGURE 3.8   The MVPL interface illustrates the concurrent data flow concept and the concurrent data flows combined using a single **Join** component (example 3.3's GCD component).

EXAMPLE 3.4: INCREMENT (CONCURRENT PROGRAM VERSION)

Explanation: Using concurrent programming design combined with the timer component to rewrite example 3.2, such that one thread updates the counter variable (**Counter**) and the other reports the number in the variable concurrently.

Skill: The skill required is the ability to build and use a timer component.

Completed diagram:

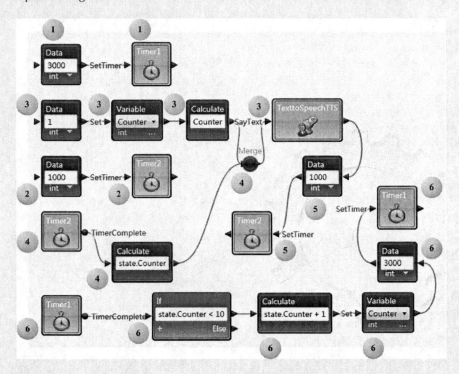

(Numbered balls represent the steps described below.)

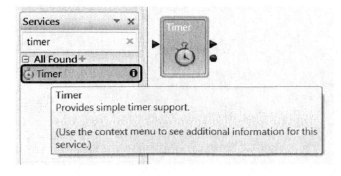

Step 1a—Create the first **Timer** component. The aim is to periodically update the counter value (i.e., the Counter variable in example 3.2).

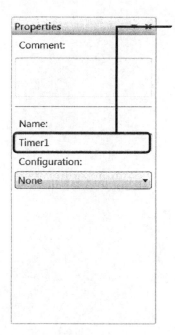

Step 1b—In Properties, change the name of the Timer component to Timer1.

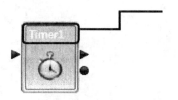

Step 1c—The name change is completed when this diagram is displayed.

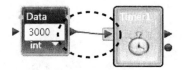

Step 1d—Use a **Data** component to set the initial value of **Timer1**.

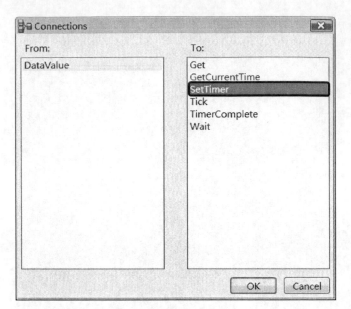

Step 1e—The connection between the **Data** and **Timer** components is set as **SetTimer**.

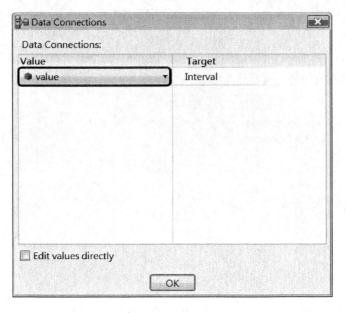

Step 1f—The **Timer** component origin is set as the **Data** component's **value**.

TIP

Timer components have a few other functions in relation to value input. For example, **GetCurrentTime** retrieves the current time and **Wait** represents waiting. You may find it interesting to experiment with each of these functions, one at a time.

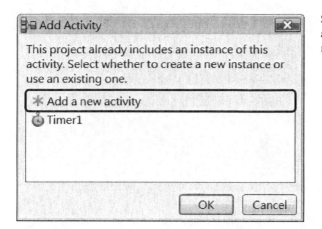

Step 2a—Create the second **Timer** component as **Timer2**. The purpose is to periodically report the counter's value.

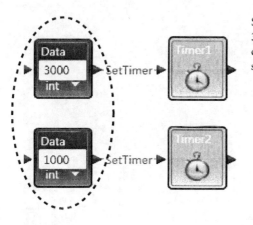

Step 2b—**Timer1**'s and **Timer2**'s initial values are set as **3000** and **1000**, respectively. As each time unit of a **Timer** component is 1 ms (0.001 second), **Timer1** is set at 3 seconds and **Timer2** is set at 1 second initially.

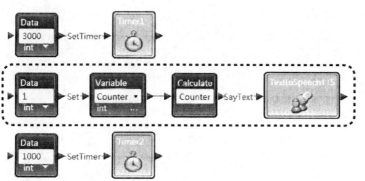

Step 3—Create a counter variable component **Variable** with the variable **Counter** in memory. Set the initial value as **1** and direct it to the voice synthesis component **Text to Speech**.

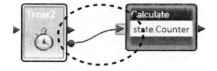

Step 4a—Create a duplicate of the **Timer2** component and direct its Notification to a **Calculate** component.

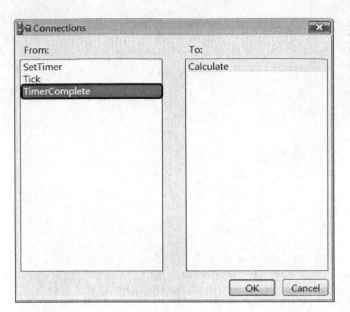

Step 4b—Set the connection between **Timer2** and **Variable** as **TimerComplete**. This triggers the **Calculate** component only when the **Timer2** component has completed timing.

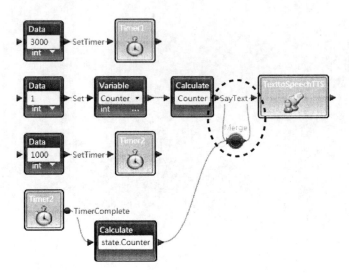

Step 4c—Connect the **Calculate** component of **Timer2** to the input end of the **Text to Speech** component, creating the **Merge** component.

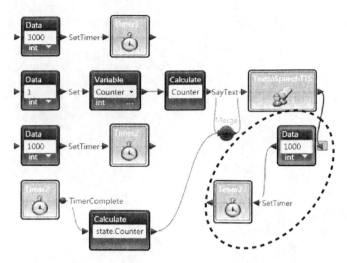

Step 5a—Create a new **Data** component to update a new **Timer2** component. After that, connect the output end of **Text to Speech** to the **Data** component.

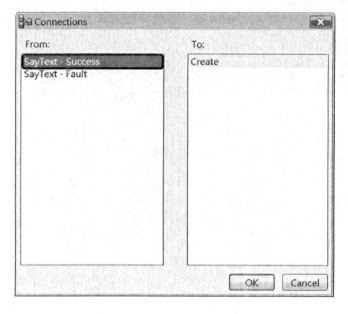

Step 5b—Set the connection between **Text to Speech** and **Data** as **SayText - Success**. This means that after **Counter**'s value has been successfully reported, the **Data** component will be triggered so that the **Timer2** component is incremented by 1 second.

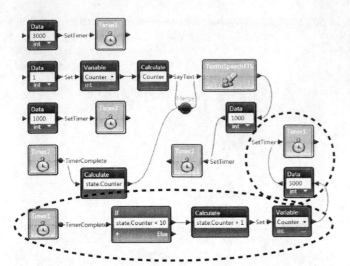

Step 6—Create the thread to start from the **Timer1** component so that it increments the **Counter** variable by 1 every 3 seconds until **Counter** is more than 10.

After completing the above steps, the example is complete. Press the <F5> key to activate the DSS service. This program will report the **Counter** value once each second (the **Timer2** component updates once every second and after that time is up, the **Counter** variable value is reported), counting from 1 to 10, so each number is reported three times (the **Timer1** component increments the **Counter** variable by 1 every 3 seconds), that is, "1, 1, 1, 2, 2, 2, …, 9, 9, 9, 10, 10, 10, 10, …" The program will not stop reporting **Counter** at 10 but rather will repeat "10" endlessly. This is because a stop condition has not been set in the program.

TIP

As the concurrent program written with MSRDS follows the DSS framework, there is no clear starting point. This is different from sequential programs. Furthermore, programs written with MSRDS have no specified end time, and even though we could modify example 3.4 to stop producing sounds, the program itself does not end. This is because the execution of the MSRDS program is handled by the DSS service and thus, as long as the DSS has not stopped, programs written with MSRDS will be held in memory.

TIP

DSS is a service and not a program. The difference between the two is that a program stops automatically at some time after execution, but a service requires other events to stop its execution (for instance, after executing the DSS service, you click the **Stop** button to halt it). You can also conceptualize a service as a program that is unable to stop by itself.

EXAMPLE 3.5: VIRTUAL SENSOR

Explanation: Use custom components to design a virtual sensor that can be used to control the orientation of a robot.

Skill: To build the notification function of a custom component and understand the use of a mathematics service component called "Math Functions."

Completed diagram:

1. Overall program

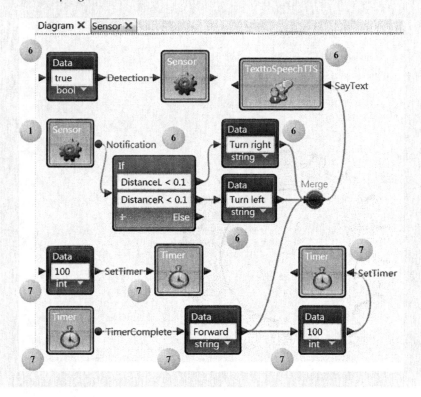

EXAMPLE 3.5 (continued): VIRTUAL SENSOR

2. Custom activity component—virtual sensor

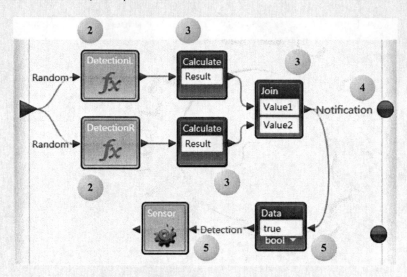

(Numbered balls represent the steps described below.)

Step 1a—Create a custom activity component and give it the name **Sensor**.

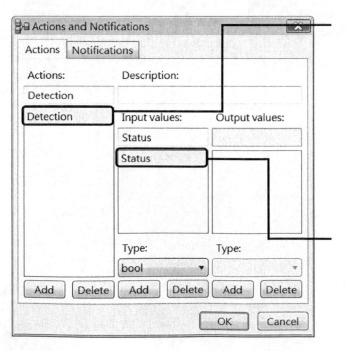

Step 1b—Add an action to the **Sensor** component called **Detection**. The **Sensor** component will detect distance when it is initiated.

Step 1c—Add the Boolean variable **Status** into the input value of the **Sensor** component, representing either the start or the end of the sensor (although this example will only use the start status).

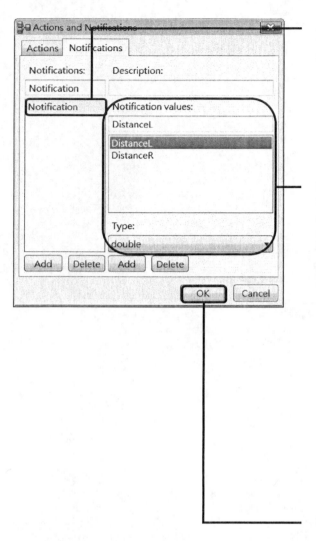

Step 1d—On the **Notifications** tab of the **Sensor** component, add a notification event and set its name as **Notification**.

Step 1e—Add two double-precision variables to the notification event, named DistanceL and DistanceR. They refer to the left and right distances detected by the Sensor component. When the values of the two variables are updated, other components related to Sensor will be informed through notification events.

Step 1f—Click OK to store these settings.

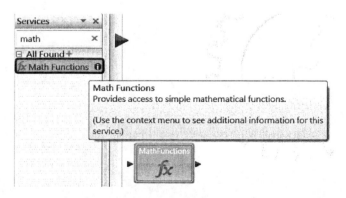

Step 2a—To program the **Sensor** component, add a **Math Functions** component. This component provides several **Math Functions**.

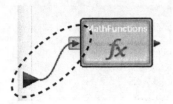

Step 2b—Create a connection between the **Sensor** component input value and the **Math Functions** component. When the **Detection** function of the **Sensor** component is started, it initiates the **Math Functions** component.

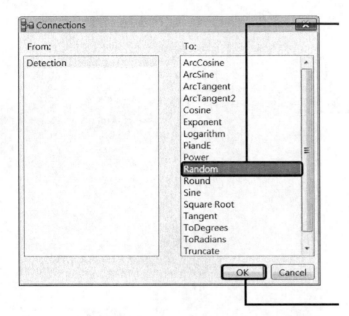

Step 2c—In the pop-up dialog box, select a function within the **Math Functions** components. Here, select **Random** to obtain a random number between 0.0 and 1.0.

Step 2d—Click **OK** to store this setting.

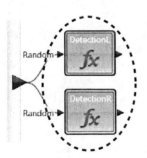

Step 2e—According to the steps described above, create two **Math Functions** components and name them **DetectionL** and **DetectionR**, representing the left and right side values of the distance detection. As we have not yet described how to combine actual sensors using MSRDS, we use functions to simulate the values of sensors.

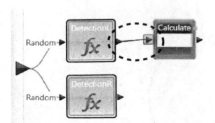

Step 3a—Create a **Calculate** component to extract the random value obtained by the **DetectionL** component.

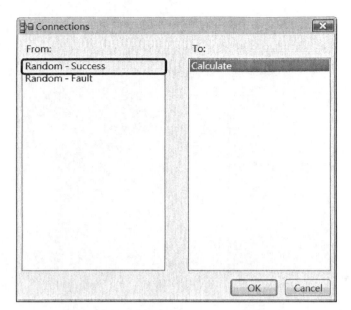

Step 3b—In the **Connections** settings dialog box, for the connection between the **Math Functions** and **Calculate** components, select **Random - Success**. This lets the **Math Functions** component initiate the **Calculate** component when it has successfully generated a random number.

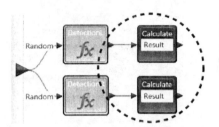

Step 3c—Within the **Calculate** component, select the default output value **Result** of the **Math Functions** component and, at the same time, create two **Calculate** components to extract the random value of the **Math Functions** component.

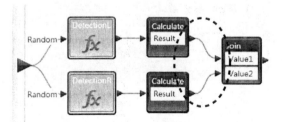

Step 3d—Create a **Join** component to combine the values of the two **Calculate** components.

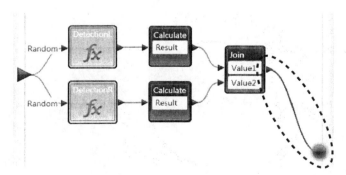

Step 4a—Create a connection between the **Join** component and the Notification of the **Sensor** component.

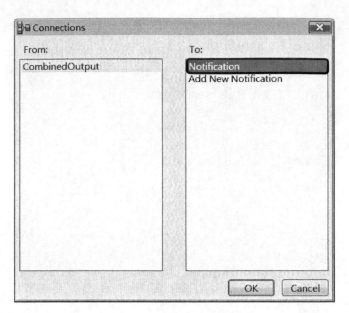

Step 4b—In the pop-up dialog box, select the custom **Notification** function.

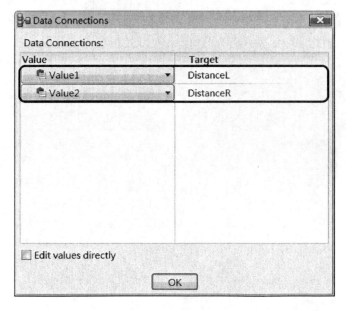

Step 4c—Separately connect the output information of the **Join** component and the variables **DistanceL** and **DistanceR** of the **Sensor** Notification.

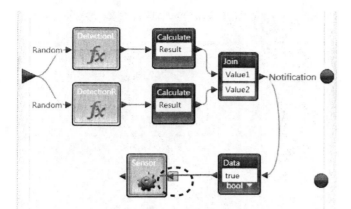

Step 5a—Now, connect the **Join** component to the new **Data** component (which triggers a Boolean value), and connect **Data** to the **Sensor** component itself, resulting in a loop.

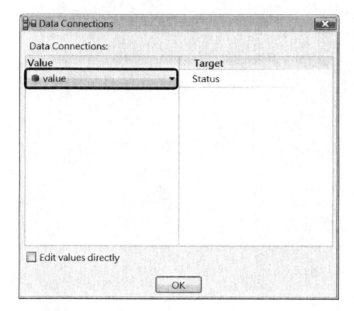

Step 5b—Set the variable connection between the **Data** and **Sensor** components.

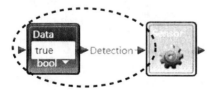

Step 6a—In the main program, add a **Data** component in front of the **Sensor** component and initiate the **Sensor** component when the Boolean value is **true**.

Step 6b—Add a **Sensor** component duplicate and connect its Notification to an **If** component.

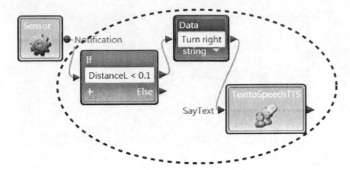

Step 6c—When the **DistanceL** value of the **If** component is less than or equal to 0.1, it means that the virtual sensor has discovered that the left distance is too close and the robot should turn to the right to avoid the obstacle. Given that the program is not connected to the robot now, let the program report **Turn right** in this event.

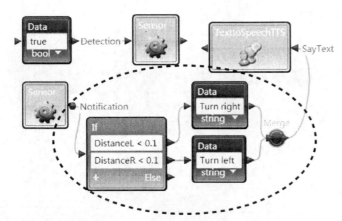

Step 6d—Similarly, set the robot's right side avoidance mechanism.

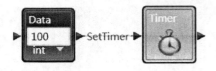

Step 7a—The robot, apart from turning left or right, should have a movement setting to move forward. This can be handled by a group of **Data** and **Timer** components.

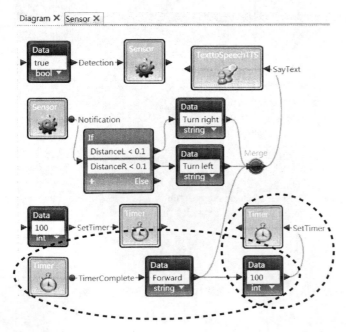

Step 7b—Set up the simulated robot so that it has the function to report **Forward** every **100** milliseconds (0.1 second). This would allow the robot to repeatedly report **Forward** to show that it is moving forward when it is turning neither left nor right. The program is now complete.

After completing the above steps, the example is complete. Press the <F5> key to start the DSS service. This example will randomly produce the words "Forward," "Turn left," and "Turn right," indicating that the simulated robot is moving forward, turning left, or turning right. This example also demonstrates the ability of the MSRDS robot program to perform concurrent processing on several sensory values. As real robots are normally equipped with several sensors, returned results can be passed to the processor through the concurrent processing threads for handling (therefore illustrating the performance of the DSS service for concurrent processing). This hides the complexity of interactions between multiple sensor devices, allowing the programmer to focus on the sensors alone in the designing of necessary behavioral reactions for the robot. In addition, this lowers the complexity of robot development.

EXAMPLE 3.6: (CHALLENGING TOPIC) ALGORITHM FOR THE GOLDEN RATIO, φ

Explanation: Use an iteration method to obtain the golden ratio, φ (pronounced as *phi*). φ is a number in mathematics with value $(1+\sqrt{5})/2$ (approximately 1.618). As the size ratio of many living creatures takes this value, it is called the golden ratio. The steps to iteratively obtain φ are as follows:

1. Take any two real nonzero numbers, a and a_1.
2. Calculate $\varphi_1 = a_1/a$.
3. Calculate $a_2 = a + a_1$.
4. Calculate $\varphi_2 = a_2/a_1$.
5. Calculate $a_3 = a_1 + a_2$.
6. Iteratively calculate $a_n = a_{n-2} + a_{n-1}$ and $\varphi_n = a_n/a_{n-1}$. As the value of n increases, φ_n will approach the real value of φ.

The steps of the iteration method are shown in the table below:

Iteration	Base Number	φ
	$a = 1$	
1	$a_1 = 2$	$\varphi_1 = a_1/a = 2$
2	$a_2 = a + a_1 = 3$	$\varphi_2 = a_2/a_1 = 1.5$
3	$a_3 = a_1 + a_2 = 5$	$\varphi_3 = a_3/a_2 = 1.666666\ldots$
4	$a_4 = a_2 + a_3 = 8$	$\varphi_4 = a_4/a_3 = 1.6$
5	$a_5 = a_3 + a_4 = 13$	$\varphi_5 = a_5/a_4 = 1.625$
6	$a_6 = a_4 + a_5 = 21$	$\varphi_6 = a_6/a_5 = 1.615384\ldots$
7	$a_7 = a_5 + a_6 = 34$	$\varphi_7 = a_7/a_6 = 1.619047\ldots$
8	$a_8 = a_6 + a_7 = 55$	$\varphi_8 = a_8/a_7 = 1.617647\ldots$
9	$a_9 = a_7 + a_8 = 89$	$\varphi_9 = a_9/a_8 = 1.618181\ldots$
10	$a_{10} = a_8 + a_9 = 144$	$\varphi_{10} = a_{10}/a_9 = 1.617977$
		$\ldots$

Skill: To use the service component Simple Dialog to achieve a dialog functionality, and to use multiple custom activity components to simplify program design to write more complex programs.

EXAMPLE 3.6 (continued): (CHALLENGING TOPIC) ALGORITHM FOR THE GOLDEN RATIO, φ

Completed diagram:

1. Overall program

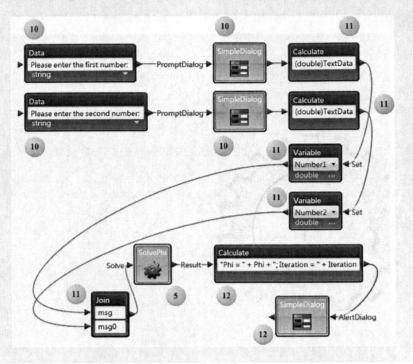

EXAMPLE 3.6 (continued): (CHALLENGING TOPIC) ALGORITHM FOR THE GOLDEN RATIO, φ

2. Custom activity component—SolvePhi

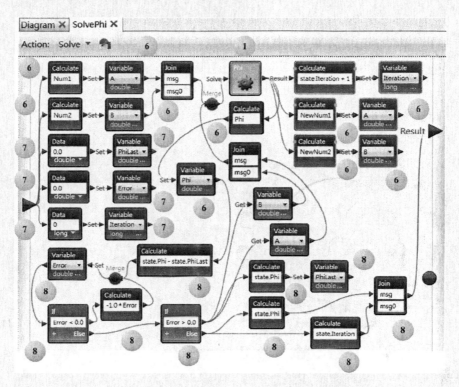

EXAMPLE 3.6 (continued): (CHALLENGING TOPIC)
ALGORITHM FOR THE GOLDEN RATIO, φ

3. Custom activity component—Phi

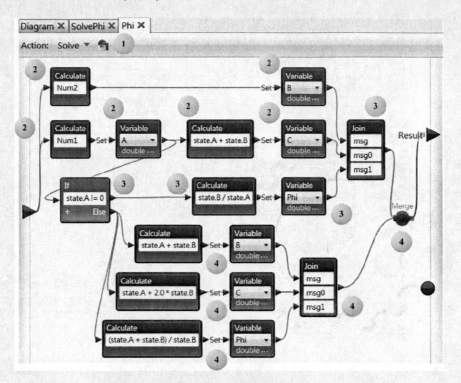

(Numbered balls represent the steps described below.)

Step 1a—Create a custom activity component and name it **Phi**. Its purpose is to solve a_n/a_{n-1} as described in the above iteration method.

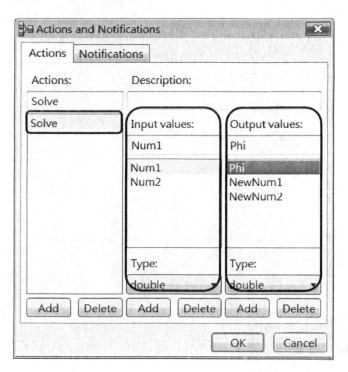

Step 1b—**Phi's** component input and output are set as follows:

Actions: Solve (solve φ).

Input values: Two real numbers **Num1** and **Num2**.

Output values: Output real number **Phi** (φ) and two updated numerical values **NewNum1** and **NewNum2**.

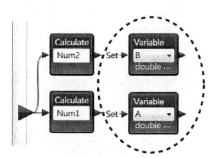

Step 2a—At the initial stage of the **Phi** component, set the two real number variables **A** and **B**.

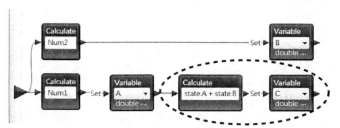

Step 2b—Calculate real number **C=A+B**, that is, $a_n = a_{n-2} + a_{n-1}$ of the iteration method.

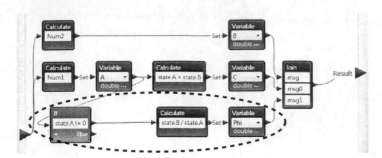

Step 3a—When **A** is nonzero, calculate the value of the real number B/A, and store the value into the real number variable **Phi**, namely $\varphi_n = a_n / a_{n-1}$ in the iteration method. Also, output the combined **B**, **C**, and **Phi**.

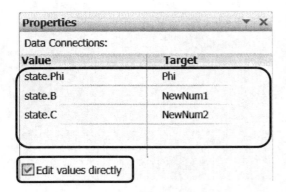

Step 3b—After the **Edit values directly** box is ticked, edit the correlation between the outputs as **state.Phi**, **state.B**, and **state.C**, corresponding to **Phi**, **NewNum1**, and **NewNum2**, respectively.

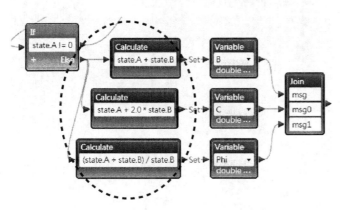

Step 4a—When **A** equals zero, use $a_n = a_{n-2} + a_{n-1}$ twice. Therefore, **B=A+B**, **C=A+2B**. At this point, recalculate the φ value.

TIP

As **A** and **B** cannot both be zero, there is no need to be concerned with the issue that **B** is zero in **(A+B)/B** when **A** is zero. You might find it interesting to include additional functionality into the **Phi** component to determine if A and B are both zero due to user input error.

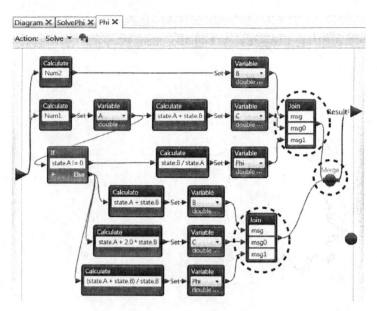

Step 4b—When **A** is equal to zero, the calculated φ value is aggregated at output. When there are two different **A** value considerations, the φ value output mode is **Merge**.

Step 5a—Create a custom activity component and name it as **SolvePhi**. Its purpose is to iteratively solve the φ value.

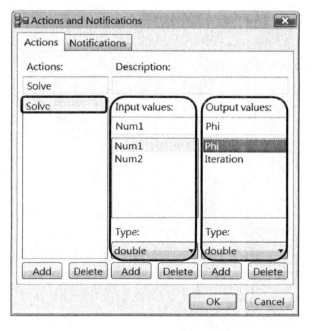

Step 5b—The **SolvePhi** component's input and output are set as follows:

Actions: Solve (repeatedly solve φ).

Input values: Two real numbers, **Num1** and **Num2**.

Output values: Output real value **Phi** (φ) and the number of iterations.

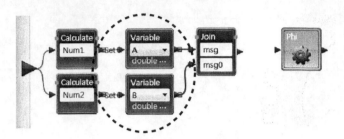

Step 6a—At the initial stage of the **SolvePhi** component, set the real number variables **A** and **B**.

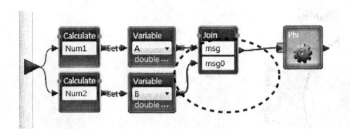

Step 6b—Connect the combined **A** and **B** to the **Phi** component.

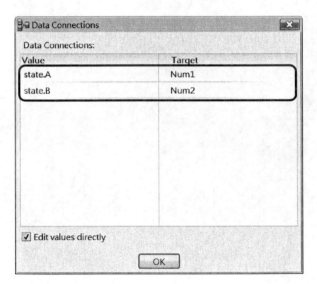

Step 6c—Enter **state.A** and **state.B** into **SolvePhi** to **Phi** component's parameters, which respectively correspond to **Num1** and **Num2** in the **Phi** component.

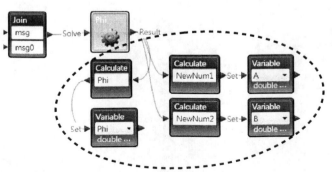

Step 6d—Separately connect the **Phi** component's output to the original real number variables **A**, **B**, and **Phi**.

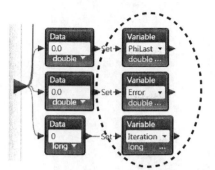

Step 7a—At the initial stage of the **SolvePhi** component, another set of three real number variables needs to be set:

PhiLast: Store the last iteration's Phi value.

Error: Store the value difference of Phi between two iterations.

Iteration: Store the number of iterations.

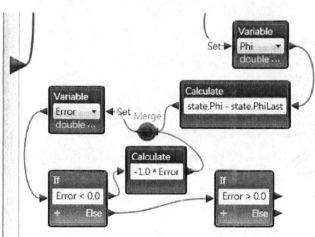

Step 7b—After the operation of the **Phi** component described in Step 6d is complete, update the **Iteration** value.

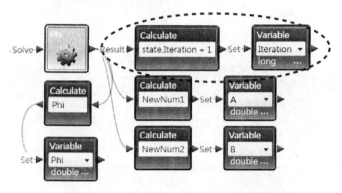

Step 8a—After obtaining the **Phi** value, calculate its difference with **PhiLast**'s value to determine if **Phi** has converged to φ's actual value.

As **Error=Phi–PhiLast** may be negative, the sign may need to be changed.

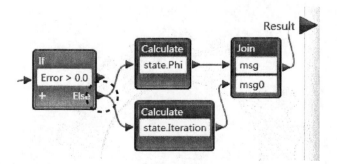

Step 8b—If the **Error** value is zero, then output **state.Phi** and the **state.Iteration** variables.

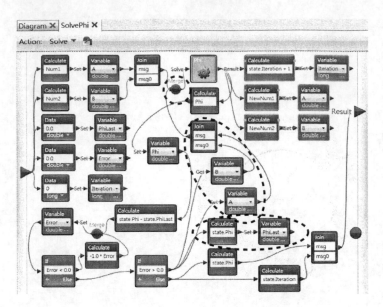

Step 8c—If the **Error** value is greater than zero, update the **PhiLast** value with **Phi**, and transfer **A** and **B** to the custom activity component **Phi** for solving the next **Phi** (φ).

Step 9—The **SolvePhi** component is now completed.

TIP

In terms of value, **Phi** and **PhiLast**'s error will not be zero, but due to the single-precision of MSRDS real numbers, the number of effective digits is less than 16. Therefore, when **Error** is less than 10^{-16}, it will be interpreted as zero, in which case it will converge.

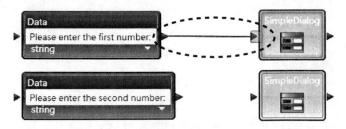

Step 10a—Return to the main program. Set two query strings **Please enter the first number:** and **Please enter the second number:** and connect them to the two **SimpleDialog** service components.

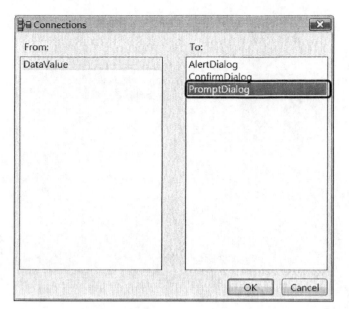

Step 10b—The connection setting between the **Data** component and the **SimpleDialog** service component is set as **PromptDialog**. This will prompt the dialog box for keyboard input.

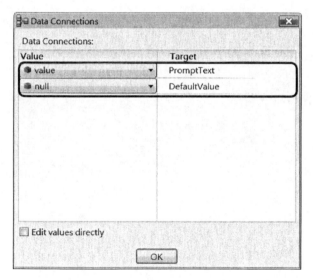

Step 10c—**PromptDialog** will start another pop-up window to set **value** to **PromptText**. Let the other item, **DefaultValue**, be **null**.

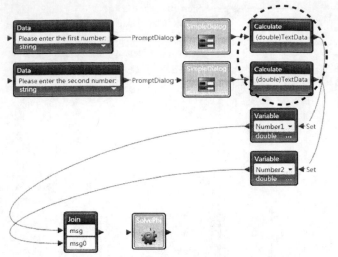

Step 11—Let the **SimpleDialog** service component's output value be transferred by the **Calculate** component (labeled **(double) TextData**) to the two variables **Number1** and **Number2**. That is, change the variables of the original string **TextData** into real numbers with the symbol "**(double)**."

Step 12a—Direct the combined value to the **SolvePhi** component.

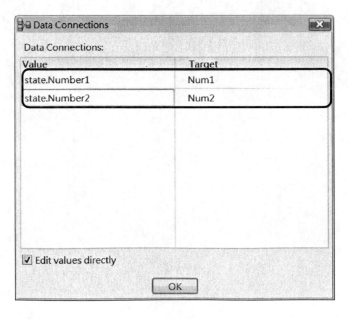

Step 12b—Ensure that the corresponding relationship is set to the directed value, that is, **Number1** and **Number2** correspond to **Num1** and **Num2**, respectively.

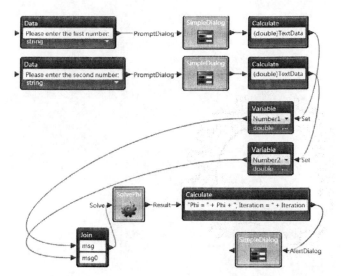

Step 13—Let the variable values **Phi** and **Iteration** outputted by the **SolvePhi** service component be directed to another **SimpleDialog** component, **AlertDialog**. The main program is now complete.

After completing the above steps, the example is complete. Press the <F5> key to start the DSS service. At runtime, two dialog boxes will pop up to query the initial two values and when these have been entered, the φ value will be calculated and the number of iterations will be shown (Figure 3.9).

3.6 CREATING A CUSTOM SERVICE COMPONENT

MSRDS provides many service components for integration with certain software and hardware, such as the *TexttoSpeechTTS* service, which allows your computer to read out words, and the *Timer* service, which reads out the system time and provides a countdown timer. These functionalities are provided by the Windows operating system, rather than by MSRDS. You will have come across these functions in the previous examples and hopefully found that they were easy to use because they require no assumed knowledge. This is the major goal of service components, as they provide an easy and straightforward way for both software and hardware to work together without the need to deal with complicated functions. In other words, MSRDS service components provide a connection between the external hardware and software.

After improving your MSRDS skills, you will need to connect with some third-party software or hardware that does not have a default setting under MSRDS. For this purpose, MSRDS allows you to create custom service components. The steps involved in creating a custom service are as follows:

- Design appropriate *state* parameters of the custom service and make it available for observation through the browser.

- Define the corresponding input, output, and notification ends of the custom service.

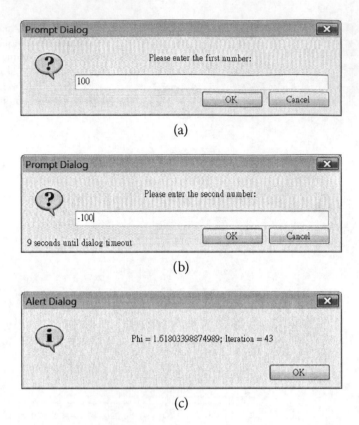

(a)

(b)

(c)

FIGURE 3.9   The execution process for example 3.6. (a) Dialog boxes will query the two initial values. The user can fill in any desired number (these two numbers cannot both be zero). (b) After the calculations are done, the φ value and the number of iterations are returned.

These can be achieved using the C# language in the *DSS Command Prompt*, which we introduced in section 2.8. You can find more detailed information from Microsoft's official teaching material, "Creating Services for Microsoft Robotics Studio," at http://www.microsoft.com/winme/0703/29490/Microsoft_Robotics_Studio_Services_Tutorials/Local/ (Figure 3.10).

3.7 EXERCISES

1. Can you think of other changes for extending the example programs in this chapter?

    a. By rewriting example 3.2, calculate the result of 1+2+3+4+5+6+7+8+9+10 and make the program read out the resulting value.

    b. By rewriting example 3.4, change **Timer1** and **Timer2**'s timing value. Check to see if there are any effects on the counting procedure.

    c. Rewriting example 3.4, can you work out how the program can stop producing sounds after reading 10 for three times repeatedly?

Figure 3.10 The teaching material for "Creating Services for Microsoft Robotics Studio" on Microsoft's official website (http://www.microsoft.com/winme/0703/29490/Microsoft_Robotics_Studio_Services_Tutorials/Local/).

    d. By rewriting example 3.5, introduce four directional sensors (front, back, left, right) in the simulated robot design. Can you see any changes in the robot's behavior?

    e. By rewriting example 3.6, read the error value from the input interface and see if it can change the number of iterations and precision.

2. In this exercise, you will develop three functions using MVPL. They will help you gain a good understanding of how to write basic loop programs and how to use a specific speech service to manipulate an audio device.

    a. Display zero to ten sequentially: you need to develop a function that can display the numbers from zero to ten in alert dialog boxes.

    b. Read odd numbers aloud: develop a function to read aloud the odd numbers between zero and ten, increasing sequentially.

    c. Lowest common multiple (LCM): design a program for calculating the lowest common multiple of any two numbers.

    d. Prime number speech: develop a function to allow the user to input a range, that is, a small number as the lower bound and a higher number as an upper bound. This function will need to find out all the prime numbers within that range and print them to an alert dialog box.

3. In the above questions, can the algorithms be written to use concurrent programming logic? Is there a difference in terms of efficiency and error reduction?

CHAPTER 4

Visual Simulation Environment

4.1 OVERVIEW

The purpose of this chapter is to use MVPL to write a robot control program in the Visual Simulation Environment (VSE). The main points covered are as follows:

- An introduction of VSE and robot simulation

- An explanation, by example, of using MVPL to operate the LEGO robots

- An explanation, by example, of hardware service components within MVPL, including the Generic Differential Drive, Desktop Joystick, and Game Controller

- An explanation, by example, of using MVPL to write an autonomous motion control program for the simulated robot, including circular motion, figure-eight-shaped motion, and spiral-shaped motion

4.2 ROBOT SIMULATION

Simulation refers to the use of a computer's computational capabilities to build a virtual world corresponding to the physical environment. It is an important link for both scientific research and industrial manufacturing. Early simulations could only be used to validate simple concepts or theories developed by scientists and engineers, but as computers have become more powerful, constructed virtual worlds have become more realistic and the applications of simulation more widespread. For a robot control system developer, computer simulation technology not only allows the design and validation of algorithms but also enables testing of interactions between robots and their environment as well as their capacities to respond to various situations.

One of the functions provided by MSRDS is the Visual Simulation Environment (VSE). In this type of simulation environment, all of the robot's body (such as the arms and sensor components), the surface on which the robot moves, the landscape observed, and the static objects connected (e.g., tables, chairs, walls, or columns in a home or hardware devices in a factory) can be constructed easily. Through VSE, the robot control system programmer does not require actual robot hardware or physically constructed objects in the robot's surrounding environment to test concepts or algorithms. VSE has the following advantages:

1. It can reduce the hardware costs for the robot development phase (especially at the early design phase). Designers can even gain experience with functionalities of existing commercial robots and can custom design unique robot components, purchasing the hardware only when the design's effectiveness has been validated.

2. The possibility to experiment gives designers a lot of flexibility, enabling them to test various ideas freely while avoiding the risk of destroying hardware setups.

3. With respect to robot education, VSE can quickly provide students, who might be lacking in knowledge of hardware, a rapid development platform, serving the purpose of promotion and training.

Nevertheless, VSE is only useful as a tool for robot education or early stage development and is unable to completely simulate environmental uncertainty. If we wish to realistically develop a stable and highly reliable robot, there is still a need to return to the actual robot's world to conduct final testing and make adjustments.

Empowered by VSE, the test environment can be constructed within MVPL. This includes the LEGO robots used in this book (which in its default state is the wheel-type robot, as shown in Figure 4.1). In the following sections, we use several examples to guide you through the use of MVPL to operate a robot in VSE.

FIGURE 4.1   The Visual Simulation Environment (VSE) provided by MSRDS.

4.3 USING MVPL TO CONTROL THE SIMULATED LEGO ROBOT

EXAMPLE 4.1: CONTROL THE SIMULATED ROBOT IN VSE

Explanation: Construct a LEGO wheel-type robot and use the built-in "Desktop Joystick" control service component to control the robot's behavior.

Skill: The ability to use the built-in "Generic Differential Drive," the "Desktop Joystick," and the "Manifest."

Completion diagram:

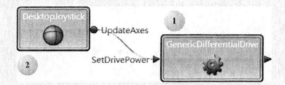

(Numbered balls represent the steps described below.)

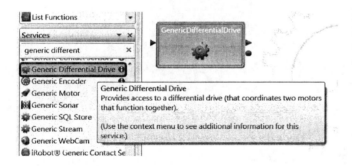

Step 1a—Create a **Generic Differential Drive** service component. This component is a common driving component.

TIP

The Generic Differential Drive is an abstract concept. It provides an integrated interface to connect different control devices and a wide range of robot hardware.

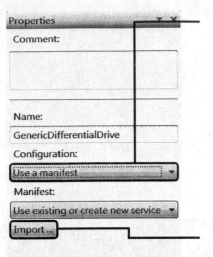

Step 1b—In the **Properties** page of the **Generic Differential Drive** service component, select the **Use a manifest** configuration.

Step 1c—Click **Import** to import the built-in manifest.

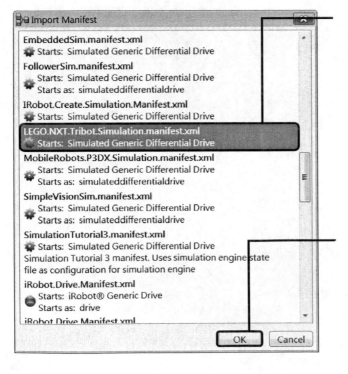

Step 1d—In the pop-up dialog box, select **LEGO.NXT.Tribot. Simulation. manifest.xml**. This is the LEGO robot built into MSRDS.

Step 1e—Click OK to accept the settings.

Step 1f—Check that the **Properties** page shows the current robot manifest.

TIP

The "manifest" is similar to an actual hardware setup. It incorporates several service components (e.g., actuators and sensors) and encapsulates functionalities of all service components. Through this manifest, designers are freed from the trouble of rebuilding each service component at every MVPL execution. MSRDS provides many common manifests for use. Chapter 5 of this book will discuss how users can create their own custom manifests.

Step 2a—Create a **Desktop Joystick** service component.

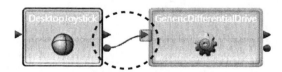

Step 2b—Connect the Notification of the **Desktop Joystick** component and the **Generic Differential Drive** service component.

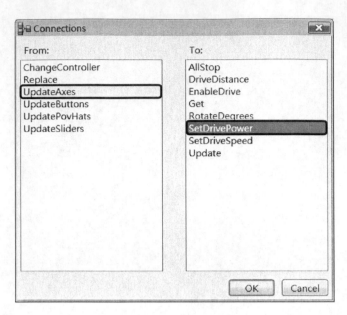

Step 2c—Make the **Connections** setting from **UpdateAxes** to **SetDrivePower**. That is, when the inner axis coordinates of the **Desktop Joystick** service component has been updated, the **Generic Differential Drive** service component's actuator will be set.

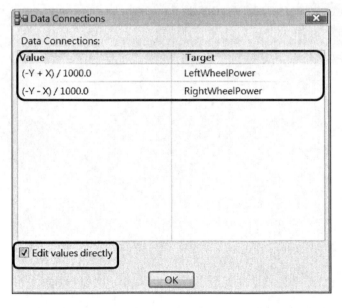

Step 2d—After the **Edit values directly** box is ticked, edit the connection values as **(-Y+X)/1000.0** to specify left wheel power, and **(-Y-X)/1000.0** to specify right wheel power. Because X and Y can be any real number, the formula should use 1000.0 and not 1000 (an integer), to prevent data type conversion errors.

TIP

As the fourth quadrant is characterized by the positive X and Y in MSRDS's coordinates (Figure 4.2), the left and right wheel settings **(-Y+X)/1000.0** and **(-Y-X)/1000.0** enable the **Desktop Joystick** service component to operate more intuitively.

1. When the controller is moved up, the coordinates are X=0 and Y<0 and the left and right wheel power value is -Y/1000.0, resulting in the robot moving forward.
2. When the controller is moved down, the coordinates are X=0 and Y>0 and the left and right wheel power value is -Y/1000.0, resulting in the robot moving backward.
3. When the controller is moved to the left, the coordinates are X<0 and Y=0; the left wheel power value is X/1000.0 and the right wheel power value is -X/1000.0; the left wheel rotates in reverse; the right wheel rotates forward, and the robot rotates anti-clockwise (left turn).
4. When the controller is moved to the right, the coordinates are X>0 and Y=0; the left wheel power value is X/1000.0 and the right wheel power value is -X/1000.0; the left wheel rotates forward; the right wheel rotates in reverse, and the robot rotates clockwise (right turn).

The above value, 1000.0, is used to adjust the movement degree of the controller and the turning speed of the robot's two wheels.

$$
\begin{array}{c|c}
(X: -, Y: -) & (X: +, Y: -) \\
\hline
(X: -, Y: +) & (X: +, Y: +)
\end{array} \longrightarrow X
$$

$$\downarrow Y$$

FIGURE 4.2   Coordinates in the MSRDS environment.

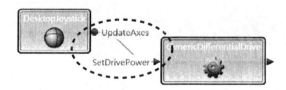

Step 2e—The connection between the **Desktop Joystick** service component and the **Generic Differential Drive** service component is now complete.

After completing the above steps, the example is complete. Press the <F5> key to start the DSS service. At runtime, the VSE environment and simulated LEGO robots will appear (Figure 4.3 to Figure 4.7). The user can use the mouse to control the **Desktop Joystick** and manipulate the simulated robot. In this environment, the **Desktop Joystick** controls the robot, and the keyboard controls the VSE environment. Pressing the <Up> and <Down> keys allows zooming into and out of the image, and dragging the screen using the mouse changes the viewing angle.

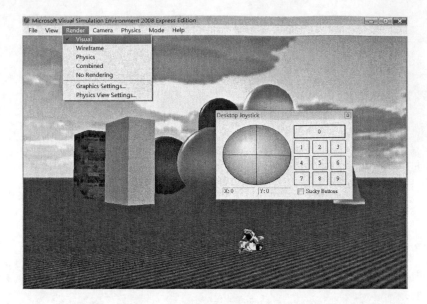

FIGURE 4.3 VSE screen: using **Desktop Joystick** to control the LEGO robots. The five rendering styles are **Visual**, **Wireframe**, **Physics**, **Combined**, and **No Rendering**. They are typically constructed separately, and the difference between each style is very important, especially in relation to collisions.

FIGURE 4.4 **Visual** render screen.

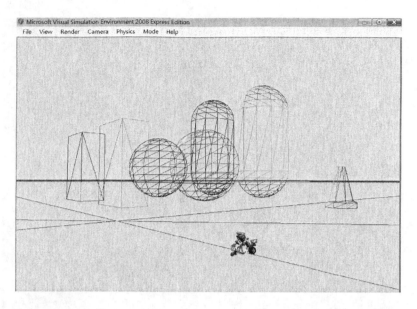

FIGURE 4.5   **Wireframe** render screen.

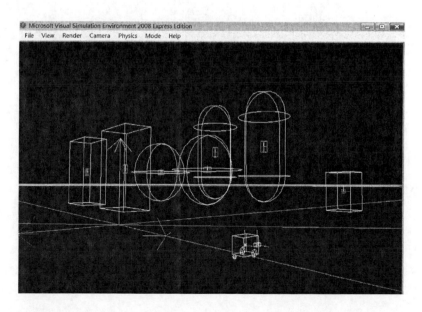

FIGURE 4.6   **Physics** render screen.

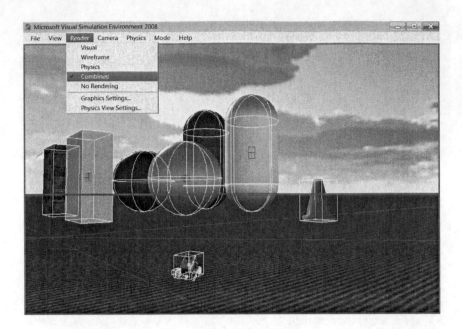

FIGURE 4.7   **Combined** render screen.

EXAMPLE 4.2: USING A GAME CONTROLLER TO CONTROL THE SIMULATED ROBOT

Explanation: Using a commercial game controller to control a robot's behavior, by rewriting example 4.1.

Skill: The ability to use the built-in game controller service component.

Completion diagram:

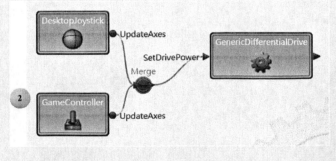

(Numbered balls represent the steps described below.)

Step 1a—Connect a commercial game controller to the computer. The controller used in this example has a USB interface.

Step 1b—Confirm that the game controller drivers have been installed. Only then can MSRDS use the controller.

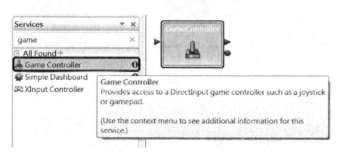

Step 2a—Create a **Game Controller** service component.

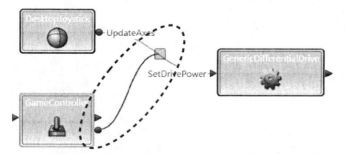

Step 2b—Connect the Notification of the **Game Controller** service component to the **Generic Differential Drive** service component.

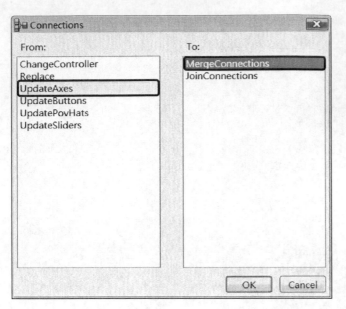

Step 2c—Select the **UpdateAxes** and **MergeConnections** options inside the **Game Controller** service component's **Connections** setting.

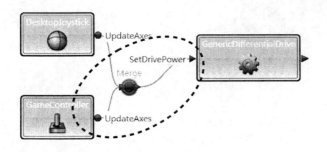

Step 2d—Complete the connection between the **Game Controller** service component and the **Generic Differential Drive** service component.

After completing the above steps, the example is complete. Press the <F5> key to start the DSS service. At runtime, the same screen will appear as in example 4.1, allowing the user to simultaneously use the **Desktop Joystick** and game controller to operate the simulated robot.

EXAMPLE 4.3: USING TOUCH SENSORS FOR OBSTACLE AVOIDANCE

Explanation: Configure the touch sensors for the robot in example 4.2.
Skill: The ability to use the built-in touch sensors' service component.
Completion diagram:

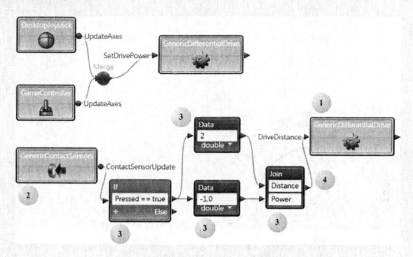

(Numbered balls represent the steps described below.)

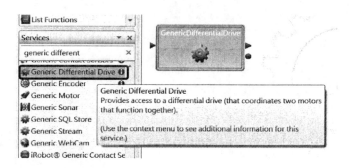

Step 1a—Add a **Generic Differential Drive** service component to example 4.2.

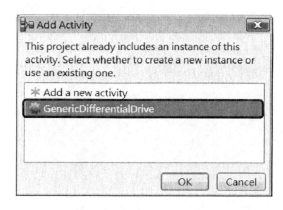

Step 1b—As this is used to control the same simulated robot (we only want to construct a copy of the existing service components), choose the existing **Generic Differential Drive** service component in the pop-up dialog box (otherwise a second robot will appear in VSE).

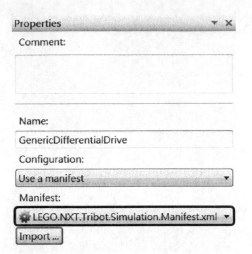

Step 1c—Confirm that the existing Generic Differential Drive is of the same manifest as that of example 4.1.

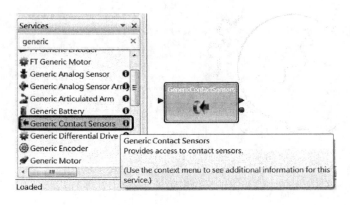

Step 2a—Create a **Generic Contact Sensors** service component.

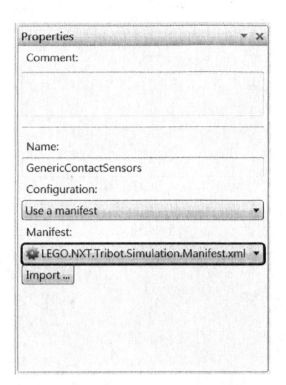

Step 2b—Set the **Properties** of the **Generic Contact Sensors** service component. The manifest needs to point to **LEGO.NXT.Tribot.Simulation. Manifest.xml**.

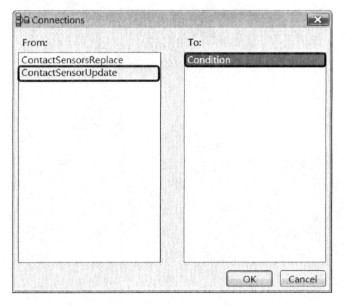

Step 3a—Connect the Notification of the **Generic Contact Sensors** service component to the **If** activity component.

Step 3b—Set the connection type between the two from **ContactSensorUpdate** to **Condition**. That is, when the sensor device data is updated, trigger the **If** activity component.

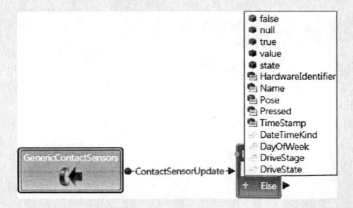

Step 3c—Set the condition of the **If** component as **Pressed == true**, meaning that when a trigger has occurred, two real number values **2.0** and **-1.0** will be created and combined with a **Join** activity component. This means that at the expected contact, the move distance is 2 and the power is –1 (causing the robot to move backward).

TIP

The condition "**Pressed == true**" can also be written as "**Pressed = true**" and has the same meaning in MVPL.

TIP

We normally would not know the service component variable names such as those used for **Generic Contact Sensors**. Nevertheless, as in the above example, upon creating the connection with the **If** activity component, left-clicking on the **condition field** of the **If** component will reveal that the **Generic Contact Sensors** service component possesses variables such as **HardwareIdentifier**, **Name**, **Pose**, **Pressed**, and **TimeStamp,** which can be used (Figure 4.8). Although we need to refer to the MSRDS manual for the usage details, this information will still allow us to correctly enter variable names without fail.

FIGURE 4.8 Variables that can be used for **Generic Contact Sensors** Notification (apart from **false**, **null**, **true**, **value**, and **state**, which are the variables/values used by the **If** activity component).

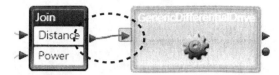

Step 4a—Connect the combined **Join** component to the **Generic Differential Drive** component service component to control the robot's behavior.

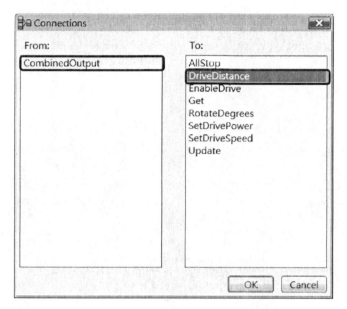

Step 4b—Set the two components' connection settings from **CombinedOutput** to **DriveDistance**, to control movement distance.

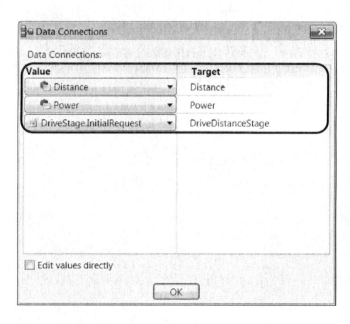

Step 4c—In the pop-up dialog box, match the **Distance** and **Power** in the two columns. However, as the variable names of the **Join** activity component (**Distance** and **Power**) are coincidentally the variable names for the **DriveDistance** action, the pop-up settings box automatically maps these variables accordingly. The third variable **DriveDistanceStage** can be set as **DriveStage.InitialRequest**. to guarantee that it works with all differential drives.

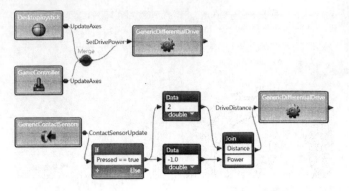

Step 5—The program design is now complete.

After completing the above steps, the example is complete. Press the <F5> key on the keyboard to start the DSS service. At runtime, the same screen in examples 4.1 and 4.2 will display, but the difference is that the touch sensor will have object avoidance capability. When the robot hits an obstacle in the environment, the robot will apply the algorithm, being 1 unit of power and moving backward 2 units in distance, to distance itself from the obstacle.

TIP

As MSRDS uses the physics engine developed by the AGEIA company (which was acquired by the well-known graphics card manufacturer NVIDIA in 2008) to construct the VSE environment, it can simulate behaviors similar to that of the real world even though it is only a simulation. The reader can experience this from experimenting with the robot's reaction in this example.

EXAMPLE 4.4: AUTONOMOUS MOTION ROBOT EXERCISE—CIRCULAR TRAJECTORY MOTION

Explanation: An intelligent robot should be able to perform independent motion, rather than being manually controlled (in examples 4.1 and 4.2 the user controls the robot's movement, while example 4.3 shows autonomous obstacle avoidance ability). This example implements a regular autonomous exercise—circular trajectory motion, which is a common task performed by robots.

Skill: The ability to use left-right wheel differentials to design an autonomous exercise.

EXAMPLE 4.4 (continued): AUTONOMOUS MOTION ROBOT EXERCISE— CIRCULAR TRAJECTORY MOTION

Completion diagram:

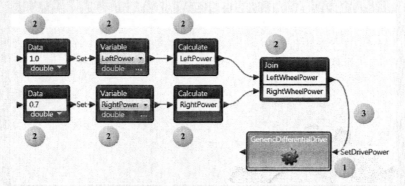

(Numbered balls represent the steps described below.)

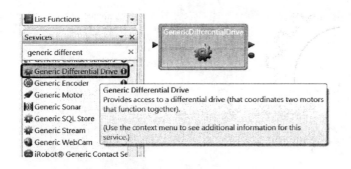

Step 1—Similarly to examples 4.1, 4.2, and 4.3, construct a **Generic Differential Drive** service component and set it to use the **LEGO.NXT.Tribot. Simulation.manifest. xml** manifest.

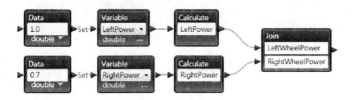

Step 2—Set a series of **Data**, **Variable**, **Calculate**, and **Join** activity components as the values for left and right wheel power, as shown in the diagram on the left.

TIP

As left wheel power (1.0) > right wheel power (0.7), when the left wheel turns quickly, the robot will be seen to move clockwise when viewed from above.

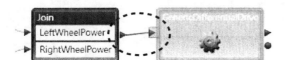

Step 3a—Create the connection between the **Join** activity component and the **Generic Differential Drive** component.

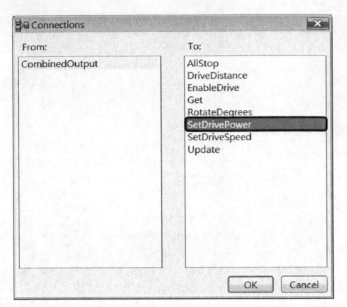

Step 3b—Set connection type as **SetDrivePower**.

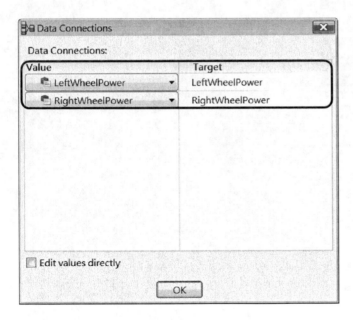

Step 3c—Set the corresponding variable relationship between the **Join** activity component and the **SetDrivePower** function, as shown in the diagram on the left.

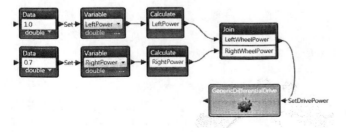

Step 4—The program design is now complete.

After completing the above steps, the example is complete. Press the <F5> key to start the DSS service. At runtime, it can be observed that the simulated robot is moving in a circular trajectory (Figure 4.9).

FIGURE 4.9   Simulated robot autonomously moving in a circular trajectory.

EXAMPLE 4.5: AUTONOMOUS ROBOT MOVEMENT— "FIGURE-EIGHT-SHAPED" MOTION

Explanation: Modify the robot to move in an "8-digit" manner (Figure 4.10), by expanding on example 4.4.

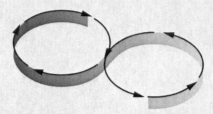

FIGURE 4.10   Autonomous robot's "figure-eight-shaped" movement trajectory.

Skill: The ability to use the Timer service component to design autonomous movement.
Completion diagram:

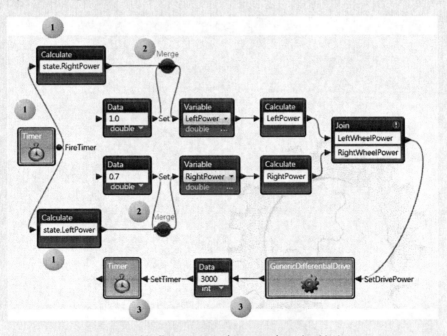

(Numbered balls represent the steps described below.)

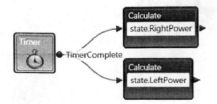

Step 1—Expanding on example 4.4, construct a **Timer** component and trigger two **Calculate** components, one calculating the **RightPower** value and the other the **LeftPower** value.

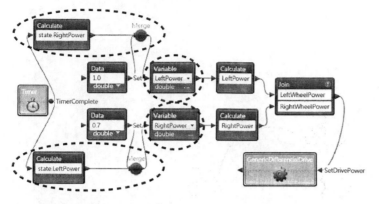

Step 2—Merge the **Timer** component's trigger values (**RightPower** and **LeftPower**) and direct them to their respective **Variables**, inverting the corresponding relationship.

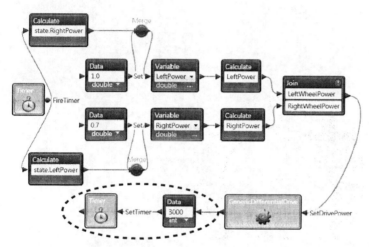

Step 3—Lastly, after the execution of the **Generic Differential Drive** component, trigger the timer procedure to set the **Timer** service component as **3000**. This will let the **Timer** service component update the robot's left and right wheel power values every 3000 ms (3 s). In this way, the robot will reverse the left and right wheel movement state. At this point, the program is complete.

TIP

In this example, the **Join** component's exclamation point means that MSRDS's compiler has given us the warning that as data flow is built upon different threads, a problem may result such that the data may not arrive when the **Join** component aggregates. However, in this example, as the route of data flow is simple and unlikely to be erroneous, we can ignore this warning.

After completing the above steps, the example is complete. Press the <F5> key to start the DSS service. At runtime, the simulated robot should move in an "figure-eight shaped" fashion.

TIP

The left-right exchange time set in this example, 3000 ms, is only a test value. In MSRDS, the built-in LEGO robot's wheel axis length with 1.0 and 0.7 wheel speed allows it to finish half the figure-eight-shaped trajectory in roughly 3 seconds, followed by a change in direction. In the future, when developing actual robots (e.g., the actual LEGO robots introduced in a later chapter), we will need to set this value only after adjustments have been made, due to differences in robot hardware design.

EXAMPLE 4.6: AUTONOMOUS ROBOT MOVEMENT— INWARD SPIRAL MOVEMENT

Explanation: Modify the robot to move in an inward spiral trajectory (Figure 4.11), by expanding on example 4.4.

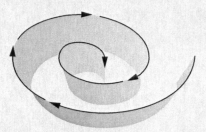

FIGURE 4.11   Autonomous robot moving in an inward spiral trajectory.

Skill: The ability to use the Timer service component to design autonomous movement.
Completion diagram:

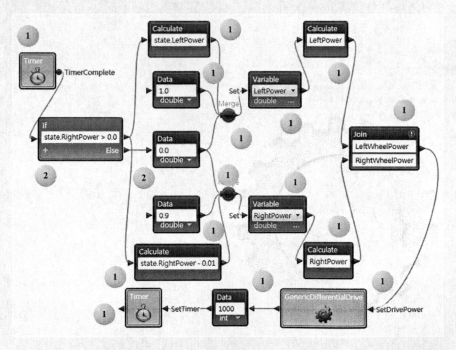

(Numbered balls represent the steps described below.)

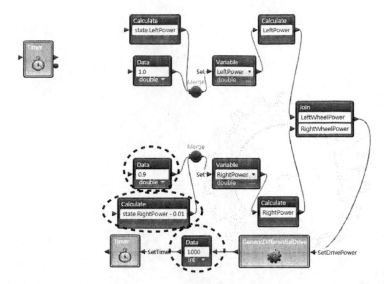

Step 1a—Modifying example 4.5, reset the time to 1000 ms (instead of 3000 ms) and clear the connection lines of the Timer component.

Step 1b—Change the original stored **state.RightPower** value in the **Calculate** activity component to **state.RightPower** - 0.01, and set the initial value of the right wheel to 0.9 to reduce the difference between the two wheel speeds.

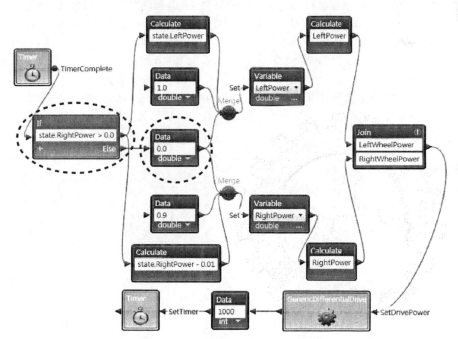

Step 2—Create an **If** activity component and a **Data** component. The program is now complete. The logic is as follows:

1. The initial wheel speed difference is 1.0 − 0.9 = 0.1, offering clockwise trajectory movement.
2. Every 1000 ms (1 s), update the value, decreasing the right wheel speed by 0.01.
3. When the right wheel speed is lower than or equal to 0.0, set the right wheel to stop (wheel speed becomes 0.0), making the robot rotate on the spot.

After completing the above steps, the example is complete. Press the <F5> key to start the DSS service. At runtime, it can be observed that the simulated robot follows an inward spiral trajectory. However, before reaching the right wheel stop state, the robot will turn upside down due to the great difference between the two wheel speeds.

TIP

As the testing is performed using the VSE environment, even though the simulated robot may turn upside down due to overly aggressive movement design, this does not result in actual hardware damage. This allows MSRDS programmers to safely test algorithms in this environment and only implement the program in the actual robot hardware after the program has been fully tested.

4.4 SCENE SETUP

VSE provides editing functionality, enabling the user to customize the scene for the virtual environment and to export it as a VSE model file format (file extension .x and .obj). If you have no prior experience in constructing 3D models, you can utilize the additional simple house model construction functionality provided by VSE, known as **FloorPlan**.

In the past, it was difficult to create complicated structures and interior simulation scenes. **FloorPlan** provides a 2D editing tool which simplifies this task. It provides objects such as walls, doors, and window items for establishing the interior of a room, house, or any other building of your choice. You can even construct a house floor plan based on a 2D house floor plan image supported by a GDI or CAD drawing. This section introduces this functionality for the reader's reference.

You can find the Visual Simulation Environment 2008 R3 item in the MSRDS program list. Here, we use the **Lego NXT Tribot Simulation** to introduce VSE's modeling function (Figure 4.12).

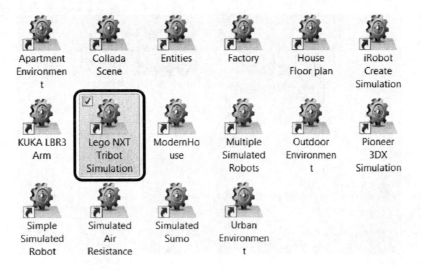

FIGURE 4.12   Visual Simulation Environment 2008 R3 item list.

Step 1a—As VSE's default execution mode is set to **Run**, change it to the editing mode **Edit**.

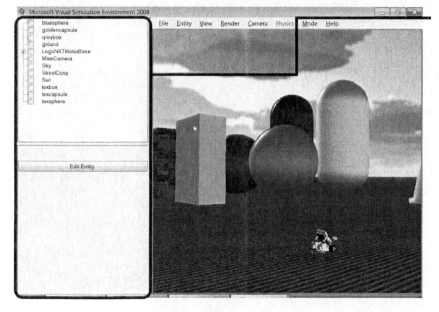

Step 1b—The **Edit** mode screen has a new **Entity** menu with various types of entities in the list on the left, including **bluesphere**, **goldencapsule**, **greybox**, **ground**, **LegoNXTBumpers**, **MainCamera**, **Sky**, **StreetCone**, **Sun**, **texbox**, **texcpsule** and **texsphere**.

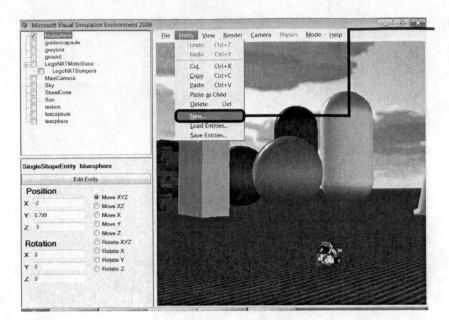

Step 2a—Select the **New** function in the **Entity** menu to add a new entity object.

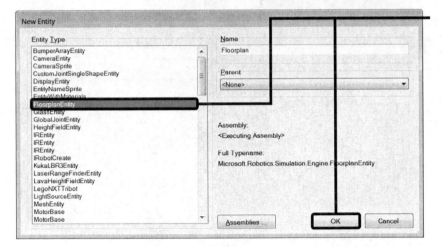

Step 2b—In the pop-up dialog box, select the **FloorplanEntity** object and select the **OK** button to confirm the new addition.

Step 2c—Following this, a parameter setup window will appear for setting the central geometry coordinates X, Y, Z for the **FloorplanEntity**. These values can be initially set to **0** in this example.

TIP

The size of VSE's popup dialog box cannot be adjusted. Therefore, a higher resolution screen is required to display all the available information. For example, the related operations for **Enter constructor parameters** cannot be displayed on a monitor with 1024×768 screen resolution.

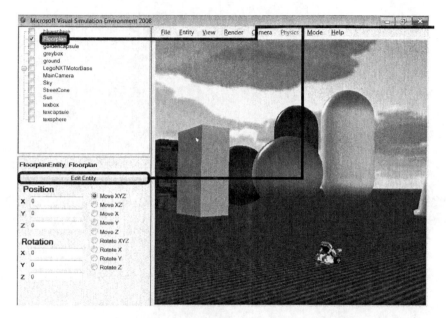

Step 3a—Returning to the main screen, the entity object **Floorplan** is shown. However, as the dimension parameters have not been set, the shape cannot be seen. Select **Edit Entity** to further edit the **Floorplan** object.

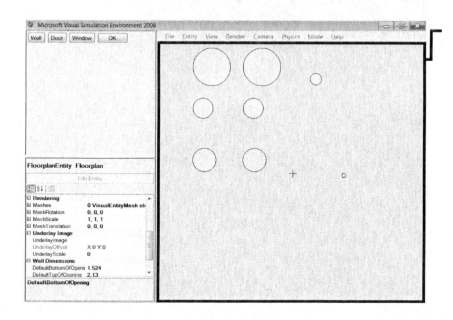

Step 3b—In the **Edit Entity** screen, all entity objects are shown on a planar view (on the right side of the screen). Dragging this screen with the right mouse button allows for zooming and adjusting the scene size, while dragging the left mouse moves the window.

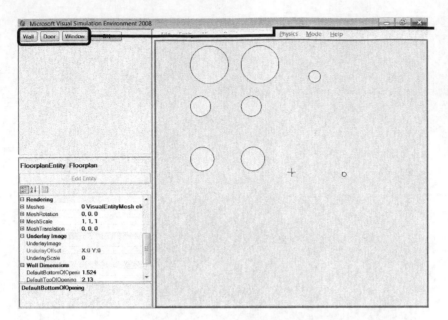

Step 3c—The left side represents the properties editing page, where **FloorplanEntity** exclusive objects such as **Wall**, **Door** and **Window** can be constructed.

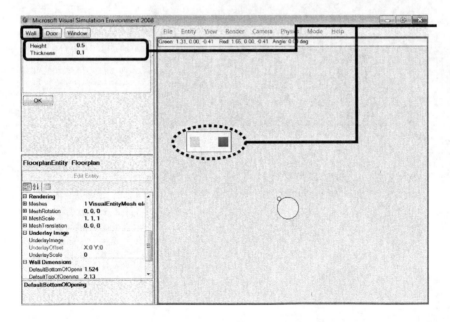

Step 3d—Select **Wall** to construct a **Wall** object with **Height** of 0.5 and **Thickness** of 0.1. At this point, the screen on the right will display the wall's position and geometric shape. Press on the left mouse button to drag the center of the wall object to change its position.

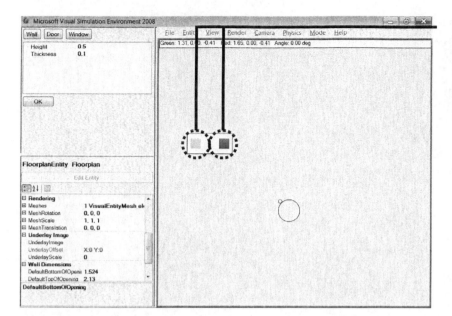

Step 3e—Press on the left mouse button to select either the green or red square on either side to adjust its length.

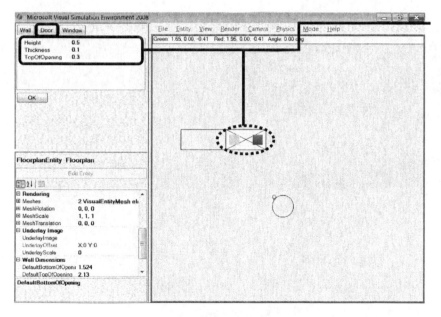

Step 3f— Select **Door** to construct a **Door** object with **Height** of 0.5, **Thickness** of 0.1, and **TopOfOpening** of 0.3. The screen on the right will display the door's position and geometric shape.

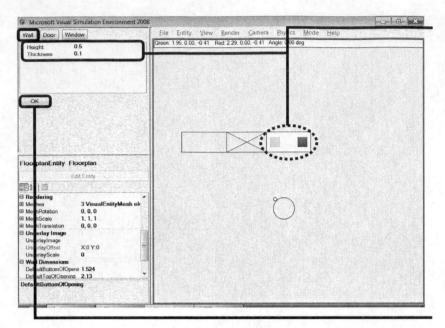

Step 3g— Select **Wall** to construct another **Wall** object beside the **Door** object. The screen on the right will display the wall's position and geometric shape.

Step 3h— Select **OK** to return to the main screen.

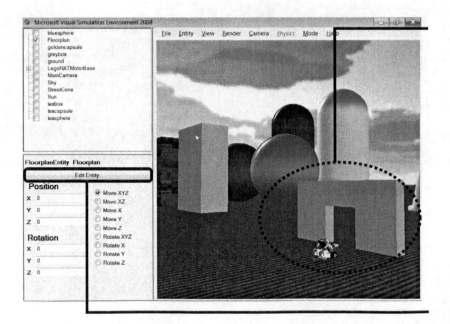

Step 4— Returning to the main screen, we can see that the wall and door objects have been constructed.

Step 5a—Select **Edit Entity** to edit the **Floorplan** object again.

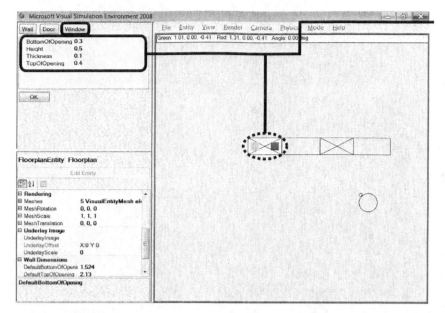

Step 5b— Select **Window** to construct a **Window** object with **Height** of 0.5, **Thickness** of 0.1, **TopOfOpening** of 0.4, and **BottomOfOpening** of 0.3. The screen on the right will display the window's position and geometric shape.

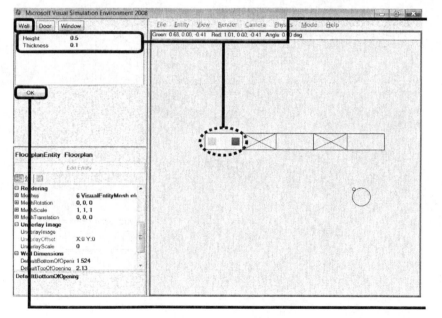

Step 5c—Select **Wall** to construct another **Wall** object beside the **Window** object. The screen on the right will display the wall's position and geometric shape.

Step 5d—Select **OK** to return to the main screen.

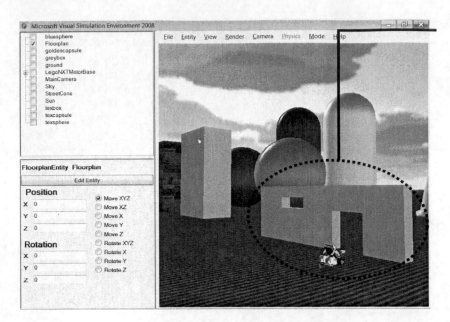

Step 6—Returning to the main screen, we can see that the wall, door and window objects have been constructed.

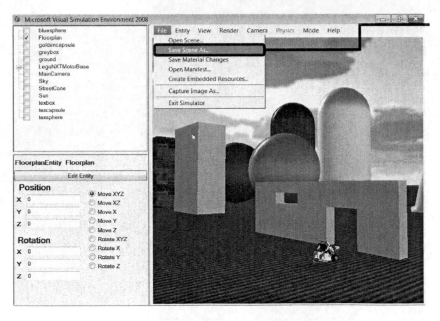

Step 7a—Select the **Save Scene As…** function in the **File** menu to save the scene you have designed.

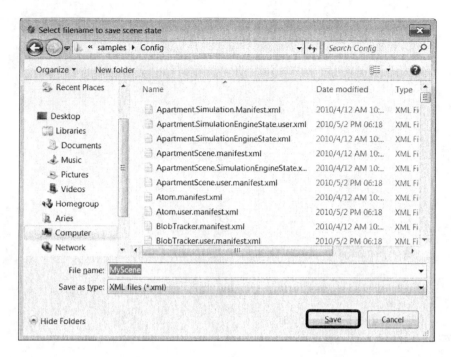

Step 7b—The save location can be set as the **\samples\Config** folder in the installation directory, storing the manifest with the existing manifests. This example will use the file name **MyLEGORobot. manifest.xml**

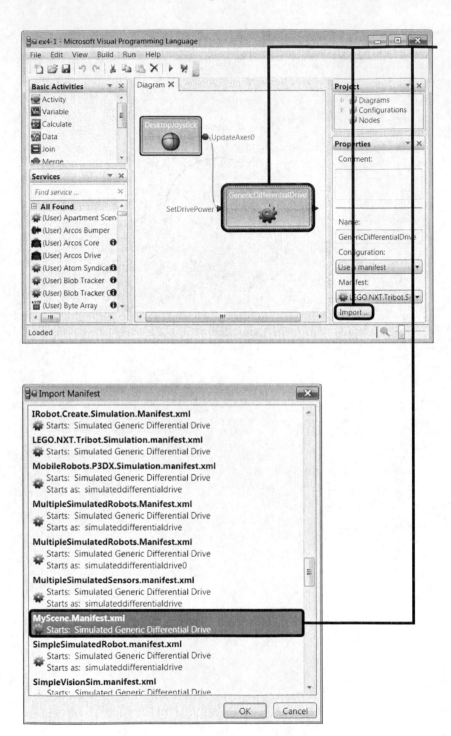

Step 8a—Currently we can test the designed scene by rewriting example 4.1. Click the **Generic Differential Drive** object to show its **Properties**, and then click the **Import** button to choose the scene we have designed in the previous steps.

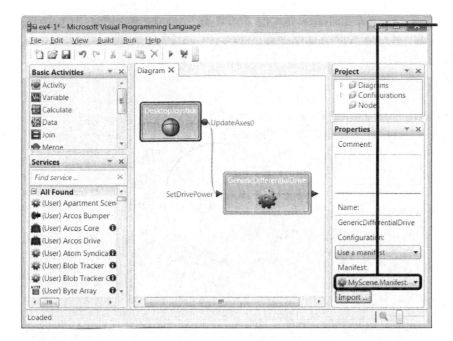

Step 8b—The **Manifest** of **Generic Differential Drive** now switches to the scene we have designed in the previous steps.

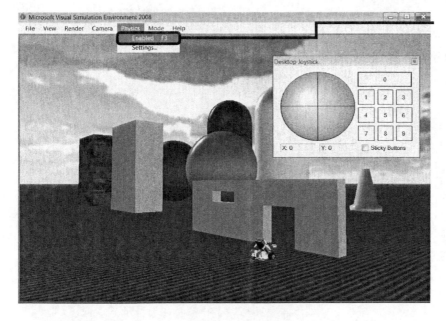

Step 8c—After completing the above steps, press the <F5> key to start the DSS service. At runtime, it can be observed that the designed scene is shown. Remember to select the **Enabled** function in the **Physics** menu to enable the physics effect before you control the simulated Lego robot.

4.5 CREATING A SIMULATED CUSTOM ROBOT IN VSE

The VSE provides a 3D simulation environment that can simulate various third-party hardware (such as the iRobot vacuum-cleaning and floor-washing robots,* LEGO educational robots, and so on) built into MSRDS. However, laboratories often have to design their own robots to meet specific requirements. Therefore, the ability to build a virtual robot in VSE at the design stage is important for reducing the cost of equipment needed. The ability to create a simulated custom robot in VSE depends on several software tools, such as the following:

- The *floorplan* editor in VSE can be used to create the simulated environment for the virtual robot (as introduced in section 4.4).
- The Microsoft XNA Game Studio (XNA), a rendering engine that can be combined and programmed as a library in MSRDS using C# language, can be used for simulating the dynamic behavior of simulated robots.
- The windows presentation foundation (WPF), a graphical subsystem for rendering user interfaces in Windows-based applications, can be used for developing a graphical user interface (GUI). It not only provides the user interface but also presents 2D and 3D drawings, fixed and on-screen documents, advanced typography, images, animation, data binding, audio, and video.

For example, in 2009 the authors developed a security robot simulator (SRS) system. We used the *floorplan* editor to create the pedestrian area for the virtual robot (Figure 4.13), the XNA to achieve the many-body dynamics of virtual robots (Figure 4.14), and the WPF to build the GUI of the control panel as a custom service in MSRDS and to visualize the virtual robot's data (Figure 4.15). This is just one of the ways of creating a simulated robot in VSE. You are not restricted to using the XNA or WPF, as other software can be used as well.

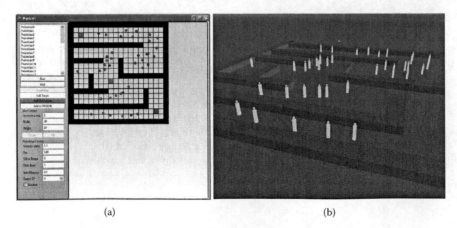

(a)   (b)

FIGURE 4.13   Screenshot of the pedestrian simulator: (a) pedestrian editor; (b) pedestrian simulation in VSE.

---

* The iRobot floor-washing robot was designed and built by the iRobot Corporation (http://www.irobot.com).

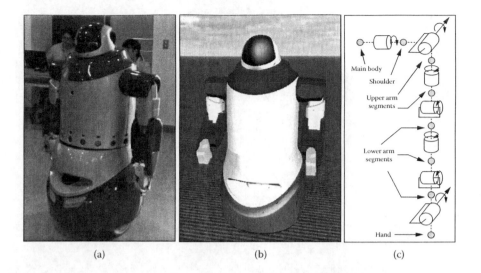

(a)   (b)   (c)

FIGURE 4.14   The intelligent security robot developed in 2009: (a) actual robot; (b) virtual security robot in VSE; (c) the articulated arm modeled and represented using many-body dynamics.

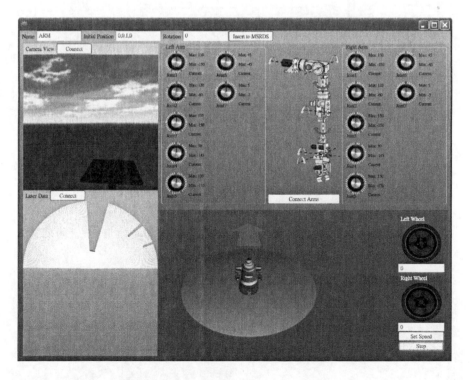

FIGURE 4.15   GUI of the developed robot unit module.

4.6 EXERCISES

1. Extending on each example of this chapter, can you think of any other changes?

   a. By rewriting example 4.1, adjust the coordination control device's **movement degree** and the robot's **drive power**. What is the result?

   b. By rewriting example 4.1, if the **coordination values** for the two wheels are different (e.g., if the **left wheel** is set as **500.0** and the **right wheel** as **1000.0**), what would be the impact?

   c. By rewriting example 4.4, if the **left and right wheel power** values are set as **1.0** and **−1.0** respectively, what will happen?

   d. By rewriting example 4.4, if the **left and right wheel power** values are set as **0.1** and **0.07** respectively, how will the robot's behavior change?

   e. By rewriting example 4.5, if the **left and right wheel exchange rates** are not 3000 ms but are set as **3500 ms** or **4000 ms**, what would be the robot's autonomous movement state?

   f. In example 4.6, what is the wheel speed difference that will cause a moving robot to turn upside down?

   g. In example 4.6, the robot turns over and the right wheel speed reduces by 0.01 every round. How is this related to resetting the time to 1000 ms in the Timer service component?

2. Experiment with robots other than the LEGO robot in each example. What is the difference?

3. Experiment with writing different programs for robot motion and program it to move in different ways. The following are some examples:

a. Triangle trajectory   b. Square trajectory

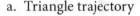
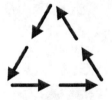
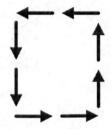

c. Oval trajectory   d. Star trajectory

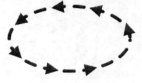

CHAPTER 5

Robot I/O Unit

5.1 OVERVIEW

Starting from this chapter, we will formally describe how a real robot operates. The purpose of this chapter is to introduce the LEGO robots' input and output components, including the Bluetooth module, sensor, motor, and so on. Through the explanation of several case examples, you will learn the skills to use these components for collaborative operations.

5.2 BLUETOOTH MODULE

The LEGO Mindstorm NXT Brick module (or in short, NXT Brick module) (Figure 5.1) is the computing core of the robot education package we are using. It encompasses a 32-bit ARM 7 microprocessor (256 kb of Flash memory and 64 kb of main memory) and an 8-bit AVR processor (24 kb of Flash memory and 512 byte of main memory). It has three inputs (for connecting to various sensors) and four outputs (for connecting to various actuator components).

The NXT Brick module supports two ways of linking to the computer and transferring commands:

1. USB 2.0—You can use the editing software provided by the LEGO robots package and the USB transfer cable to download the computer program to the NXT Brick module, allowing the NXT Brick module to autonomously control the robot.

2. Bluetooth 2.0 wireless transfer—This is the connection type used by MSRDS. The program operates and performs calculations on the computer, with the related commands being transferred via Bluetooth to the NXT Brick module to control the robot.

This book uses the Bluetooth wireless connection in teaching MSRDS programs.

FIGURE 5.1   LEGO Mindstorm NXT Brick.

EXAMPLE 5.1: LEGO MINDSTORM NXT BRICK MODULE BLUETOOTH CONNECTION SETUP

Explanation: Using Bluetooth to connect to the LEGO Mindstorm NXT Brick module.

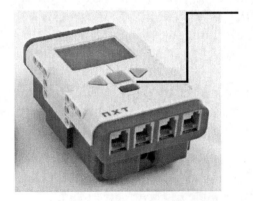

Step 1—Press the **middle rectangular button** on the NXT Brick module.

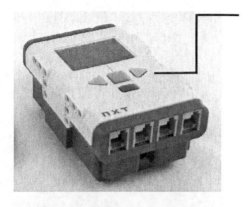

Step 2—Press the **triangular arrow button** on the NXT Brick module repeatedly until the **Bluetooth** interface is found.

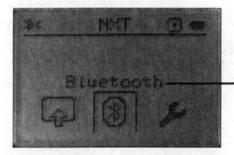

Step 3a—Press the **orange button** on the NXT Brick and enter the **Bluetooth** menu. Then press the **triangular arrow button** on the NXT Brick again until the **Connections** menu option is found.

Step 3b—Choose **any number group**. (The NXT Brick module has four numbers for use).

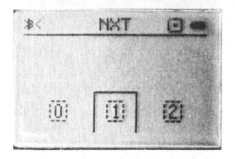

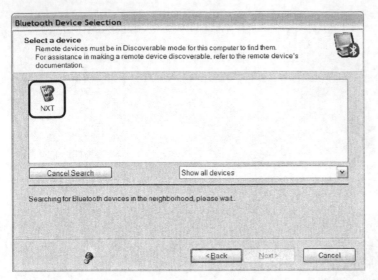

Step 4—Start the **Bluetooth functionality** on the computer. Search for the NXT Brick device; its default name is **NXT**.

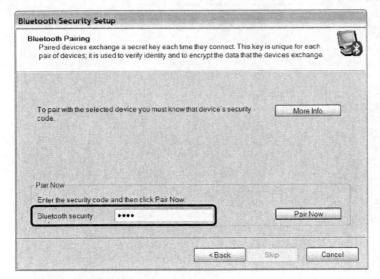

Step 5—Enter **any password** for the Bluetooth communication key. For this example, **1234** is used.

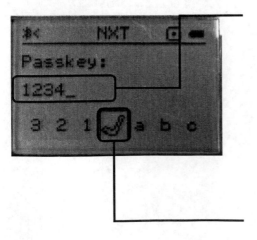

Step 6a—At this time, the NXT Brick should produce a short "beep" sound. The screen will then ask for a passkey. Please fill in the **password** from Step 5. (In this example, it would be **1234**.)

Step 6b—Choose the tick to confirm.

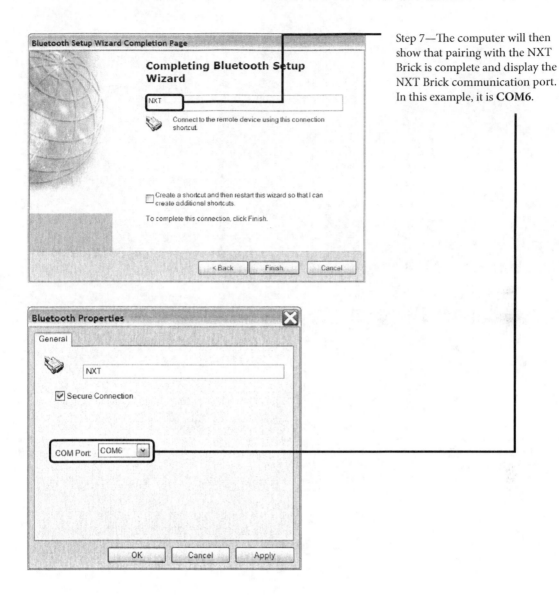

Step 7—The computer will then show that pairing with the NXT Brick is complete and display the NXT Brick communication port. In this example, it is **COM6**.

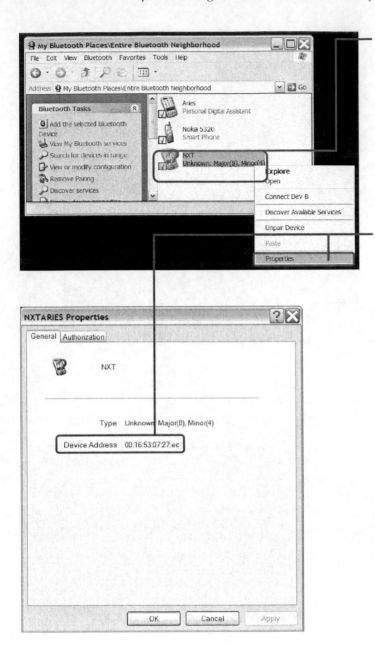

Step 8a—In the computer's Bluetooth devices page, the paired NXT Brick module can be found and renamed.

Step 8b—Click **Properties** to view hardware details such as its MAC address. Since the MAC address is unique, it is useful for ensuring that you have connected to the correct LEGO NXT Brick in a classroom with multiple Bluetooth devices.

TIP

If you are a teacher using multiple LEGO NXT Bricks in a classroom, it is suggested that you change the names of each NXT Brick with LEGO's default program (LEGO NXT 2.0 Programming) before use. Otherwise, you can only identify the corresponding LEGO NXT Brick with the MAC address.

Through example 5.1, the computer completes the Bluetooth wireless connection with the NXT Brick module. After this, when MSRDS is in operation, the related computing work is handled by the computer's processor while the NXT Brick module receives real-time commands and connects with components such as sensors and actuators. This has the advantage of improving the robot's calculation efficiency but is restricted by a limited Bluetooth communication range (generally less than 10 meters, and as determined by the computer and NXT Brick's Bluetooth hardware features).

5.3 SENSOR AND MOTOR

In Chapter 4, we described how to operate the touch sensor. In this chapter, we introduce the real robot's sensors such as the touch sensor, light sensor, and ultrasonic sensor as well as their collaborative operation with the motor.

EXAMPLE 5.2: FLASHLIGHT

Explanation: Using touch sensor and light sensor to implement a flashlight.
Skill: The ability to operate the real robot's touch sensor and light sensor devices in the MVPL environment.
Completion diagram:

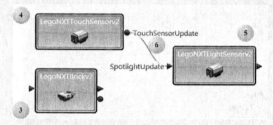

(Numbered balls represent the steps described below.)

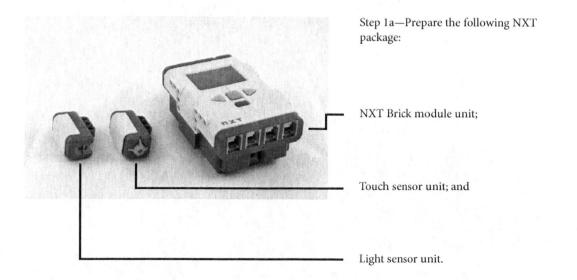

Step 1a—Prepare the following NXT package:

NXT Brick module unit;

Touch sensor unit; and

Light sensor unit.

Step 1b—Connect the sensor's cable.

Step 1c—The sensor connection is complete.

Step 2a—Connect the other end of the sensor cable to the NXT Brick module.

Step 2b—The connection between NXT Brick module and sensor is now complete.

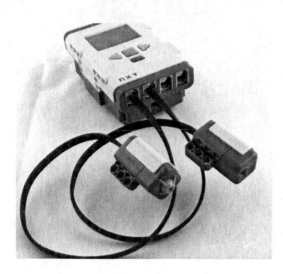

Step 2c—This example connects the touch sensor to NXT Brick module's **Sensor 1** access port and the light sensor to NXT Brick module's **Sensor 2** access port.

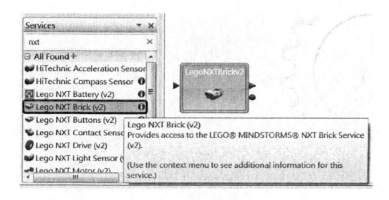

Step 3a—Create a **LEGO NXT Brick** service component. This is the service component for the actual NXT Brick Module.

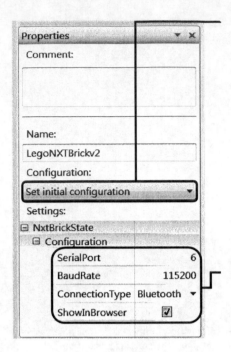

Step 3b—Set the initial state of the LEGO NXT Brick service component as **Set initial configuration**.

Step 3c—Set the connection type as **Bluetooth** and the **SerialPort** as **6** (this is the previous example's output COM port, which changes according to different computer configurations). If **ShowInBrowser** is ticked, the browser will be displayed to show the robot's operation status during program execution.

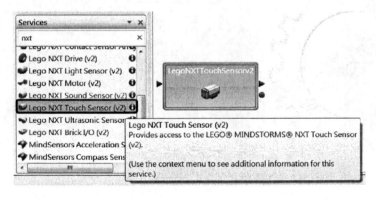

Step 4a—Create a **LEGO NXT Touch Sensor** service component. This is the corresponding service component for the real touch sensor.

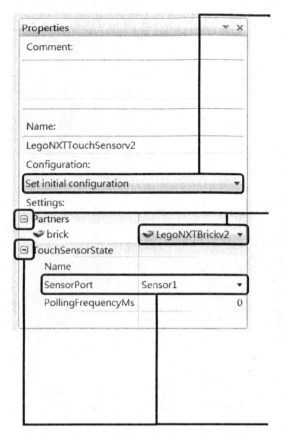

Step 4b—Set the initial configuration of the LEGO NXT Touch Sensor service component as **Set initial configuration**.

Step 4c—Select the "+" icon to open the settings page (when the "+" icon is clicked, it turns into a "–" icon and the advanced options are revealed) and set the connection component as the LEGO NXT Brick service component.

Step 4d—Select the "+" icon to open settings page and set the connection port as NXT Brick's **Sensor1** (This varies depending on your own NXT package build).

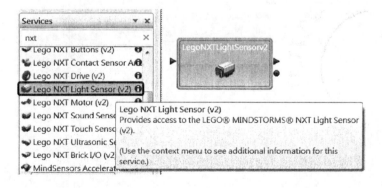

Step 5a—Create a **LEGO NXT Light Sensor** service component. This is the corresponding service component for the real light sensor.

TIP

The NXT light sensor can detect light intensity in the surroundings by emitting red light to detect the darkness value of the surroundings. The value ranges between 0 (darkest) and 100 (brightest).

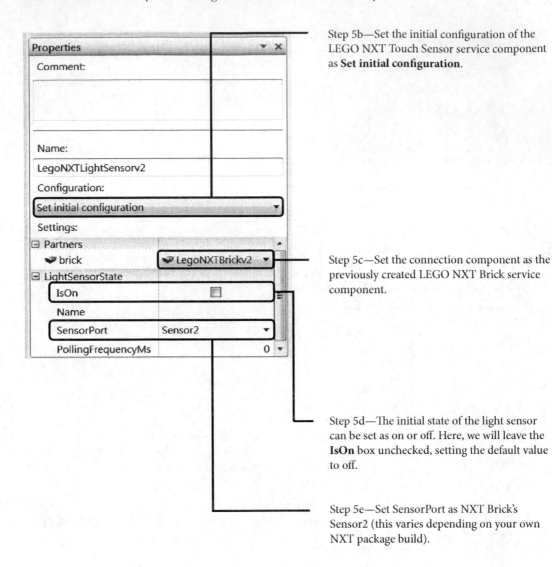

Step 5b—Set the initial configuration of the LEGO NXT Touch Sensor service component as **Set initial configuration**.

Step 5c—Set the connection component as the previously created LEGO NXT Brick service component.

Step 5d—The initial state of the light sensor can be set as on or off. Here, we will leave the **IsOn** box unchecked, setting the default value to off.

Step 5e—Set SensorPort as NXT Brick's Sensor2 (this varies depending on your own NXT package build).

Step 6a—Connect the Notification of the **LEGO NXT Touch Sensor** service component to the **LEGO NXT Light Sensor** service component.

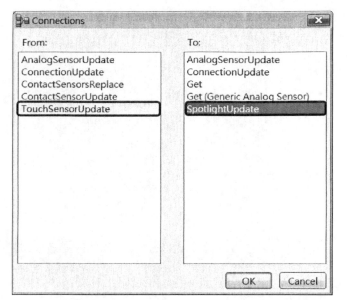

Step 6b—The connection is from **TouchSensorUpdate** to **SpotlightUpdate**, which means that when the touch sensor state is updated, the light sensor state is also updated.

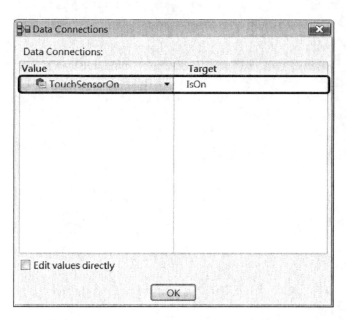

Step 6c—Configure the settings page such that when the touch sensor is in the **On** state, the **light sensor** will also be **On**.

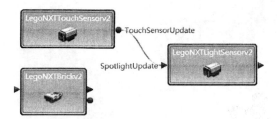

Step 7—The program design is now complete.

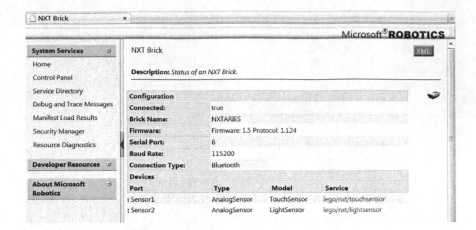

FIGURE 5.2 Using the browser to monitor the state of the LEGO Mindstorm NXT Brick. At this stage, we can click the hyperlinks of each sensor to monitor their current states. The corresponding pages are dynamic, which is useful for program debugging and development.

During execution, this program needs to turn on the actual NXT Brick module's power and Bluetooth service before executing the MVPL program (press the <F5> key to start DSS service). If everything is set up correctly, the NXT Brick module will produce a "beep" sound to show that it has successfully connected with the computer. At this point, you can press on the touch sensor and the light sensor red light will either start or stop. Apart from this, as **ShowInBrowser** in Step 3c has been ticked, the program will display the browser to monitor the actual NXT system's state (Figure 5.2).

EXAMPLE 5.3: MANUALLY CONTROLLED MOTOR

Explanation: Use the motor in place of the light sensor, creating a manually controlled motor, by rewriting example 5.2.

Skill: The ability to operate the real robot's touch sensor and motor in the MVPL environment.

Completion diagram:

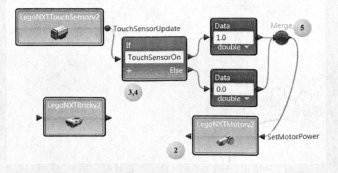

(Numbered balls represent the steps described below.)

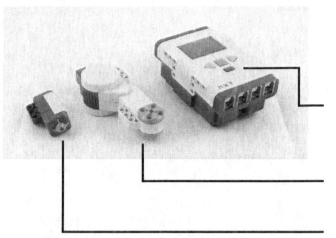

Step 1a—Prepare the NXT package below:

NXT Brick module unit;

Motor unit; and

Touch sensor unit.

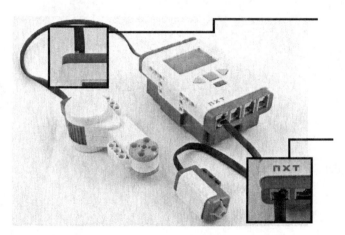

Step 1b—Connect the motor to NXT Brick module's **MotorA** port.

Step 1c—Connect the touch sensor to NXT Brick module's **Sensor1** port.

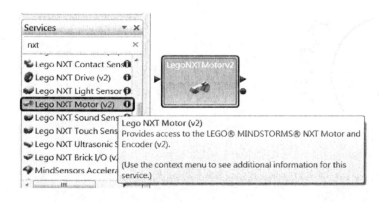

Step 2a—Modifying example 5.1, create a **LEGO NXT Motor** service component. This is the service component corresponding to the actual motor.

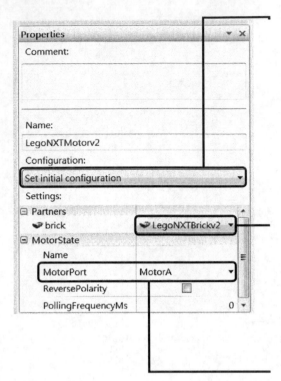

Step 2b—Set the initial configuration of the LEGO NXT Touch Sensor service component as **Set initial configuration**.

Step 2c—Set the connection component as the previously created LEGO NXT Brick service component.

Step 2d—Set the connection port as NXT Brick's MotorA (this will vary according to your own NXT package build).

Step 3a—Connect the Notification of the LEGO NXT Touch Sensor service component to an **If** activity component.

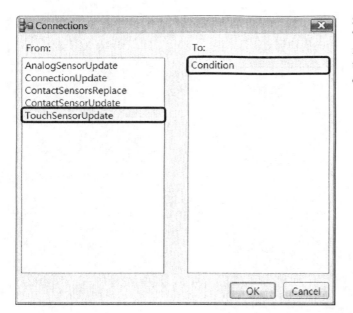

Step 3b—Set the connection from **TouchSensorUpdate** to **Condition**, meaning that when the state of the touch sensor is updated, the **If** activity component will be triggered.

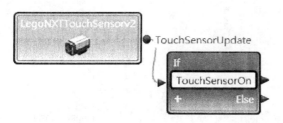

Step 4a—Set the **If** activity component's condition as **TouchSensorOn**.

TIP

As **TouchSensorOn** is LEGO NXT Touch sensor's in-built Boolean variable, there is no need to write it as TouchSensorOn == true.

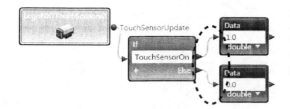

Step 4b—Connect the results of the **If** activity component to real values **1.0** (true) and **0.0** (false). When the touch sensor has been pressed down, the value **1.0** will be sent; otherwise the value **0.0** will be sent.

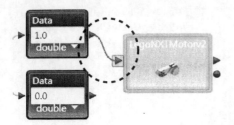

Step 5a—Separately establish connections between the two real values and the **LEGO NXT Motor** service component.

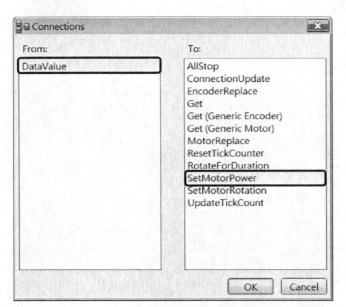

Step 5b—Set the connection from **DataValue** to **SetMotorPower**, meaning that the data value is used to set the motor power.

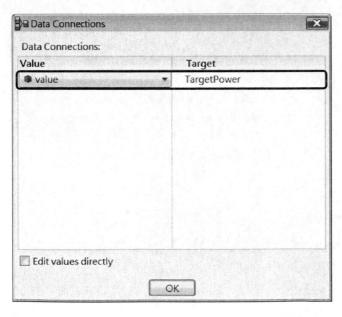

Step 5c—Set the variable corresponding relationship so that the **value** corresponds to **TargetPower**.

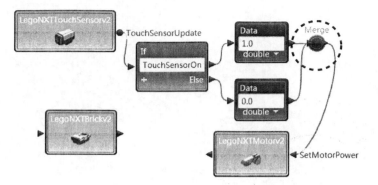

Step 6—**Merge** the two **Data** activity components **1.0** and **0.0** and connect it to the **LEGO NXT Motor** service component. The program design is now complete.

Similarly to example 5.2, during execution, this program needs to turn on the actual NXT Brick module's power and Bluetooth before executing the MVPL program (press the <F5> key to start DSS service). If everything has been set up correctly, the NXT Brick module will produce a "beep" sound to show that it has successfully connected with the computer. At this point, you can press on the touch sensor and the motor will either perform revolutions or stop.

EXAMPLE 5.4: LIGHT-CONTROLLED MOTOR

Explanation: Use the light sensor in place of the touch sensor to create a light-controlled motor, by rewriting example 5.2.
Skill: The ability to operate the real robot's light sensor and motor in the MVPL environment.
Completion diagram:

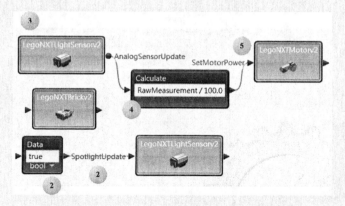

(Numbered balls represent the steps described below.)

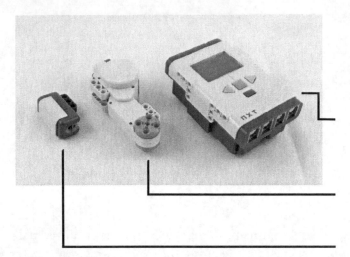

Step 1a—Prepare the NXT as below:

NXT Brick Module unit;

Motor unit; and

Light sensor module unit.

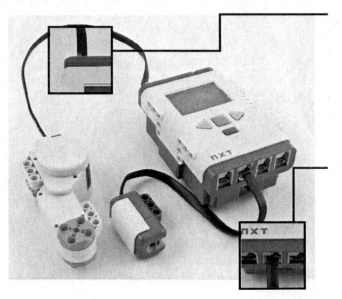

Step 1b—Connect the motor to the NXT Brick module's **MotorA** port.

Step 1c—Connect the light sensor to the NXT Brick module's **Sensor2** port.

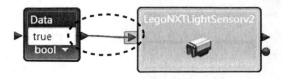

Step 2a—Modifying example 5.3, set **LEGO NXT Motor's** connection port as **MotorA** and **LEGO NXT Light Sensor** service component's connection port to **Sensor2**. After this step, create a **Data** activity component, which stores the **true** Boolean value, and connect this to the **LEGO NXT Light Sensor** service component. This is to start the light sensor's automatic detection.

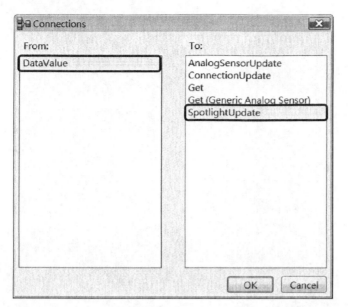

Step 2b—Set the connection from **DataValue** to **SpotlightUpdate**, meaning that the light sensor will be configured first at program initialization.

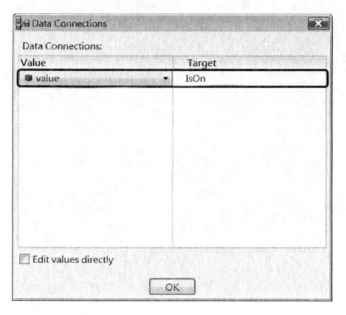

Step 2c—Set the corresponding variable relationship such that **value** corresponds to **IsOn**, that is, to start the light sensor device.

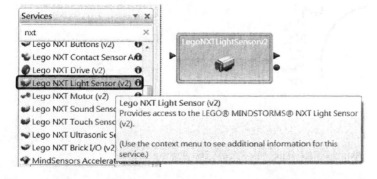

Step 3a—Create another **LEGO NXT Light Sensor** Service component.

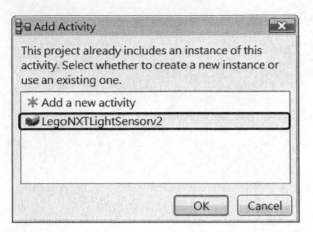

Step 3b—As there already exists a **LEGO NXT Light Sensor** service component and we only want to create a duplicate, we can select the existing component.

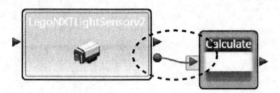

Step 4a—Connect the Notification of the LEGO NXT Light Sensor service component to a **Calculate** activity component.

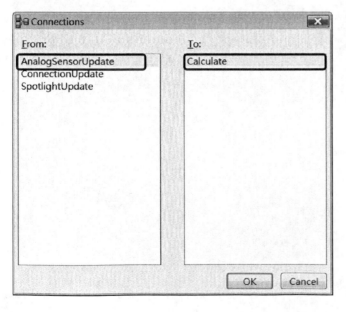

Step 4b—Set the connection from **AnalogSensorUpdate** to **Calculate**. That is, output the light sensor value of the passive light sensor to the **Calculate** activity component in analog format.

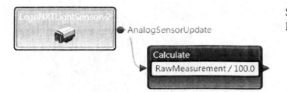

Step 5a—In the **Calculate** activity component, input **RawMeasurement/100.0**.

TIP

As the LEGO NXT Light Sensor service component's detection value ranges from 0.0 (darkest) to 100.0 (brightest), dividing by 100 will normalize the value to between 0.0 and 1.0.

Step 5b—Connect the **Calculate** activity component to the **LEGO NXT Motor** service component.

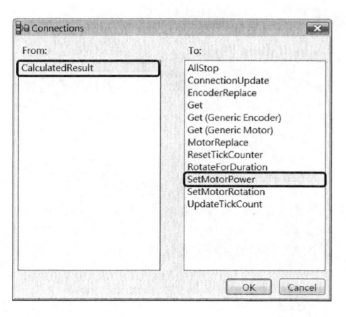

Step 5c—Set the connection from **CalculatedResult** to **SetMotorPower**; that is, use the calculated result to set motor power.

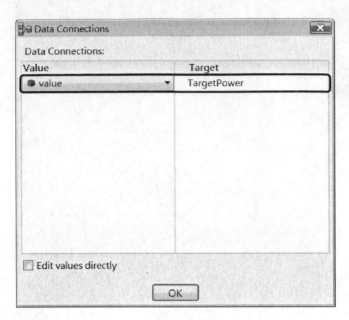

Step 5d—Set the corresponding variable relationship such that **value** corresponds to **TargetPower**. This is to ensure that the normalized light sensor value corresponds to the motor power (the two ranges from 0.0 to 1.0).

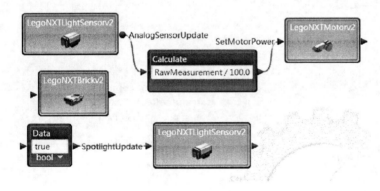

Step 6—The program design is now complete.

After the program has started, the NXT Brick module will produce a "beep" sound to show that it has successfully connected with the computer. At this point, the motor will automatically adjust its turning speed according to the light intensity. In other words, the turning speed is proportional to the light intensity.

TIP

This example covers two types of detection methods, that is, active detection and passive detection (Figure 5.3). In the dark, as the passive detection return value is zero, the environment data cannot be obtained based on passive detection alone. Instead, the active detection's red light provides light to control the environment's brightness level.

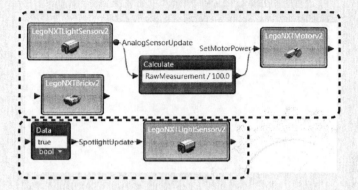

FIGURE 5.3   Example 5.4's active detection (A) and passive detection (B).

EXAMPLE 5.5: USING AN ULTRASONIC SENSOR TO CONTROL THE MOTOR

Explanation: Replace the light sensor in example 5.4 with an ultrasonic sensor, creating a distance detection controlled motor.

Skill: The ability to operate the real robot's ultrasonic sensor and motor in the MVPL environment.

Completion diagram:

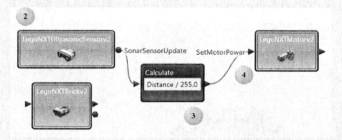

(Numbered balls represent the steps described below.)

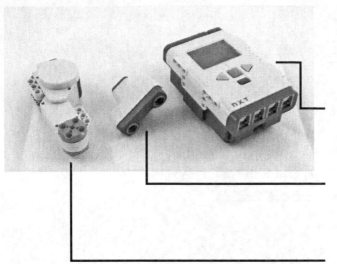

Step 1a—Prepare the NXT package below:

NXT Brick module unit;

Ultrasonic sensor module unit; and

Motor unit.

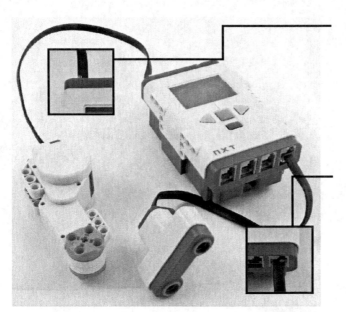

Step 1b—Connect the motor to the NXT Brick module's **MotorA** port.

Step 1c—Connect the **ultrasonic sensor** to the NXT Brick module's **Sensor4** port.

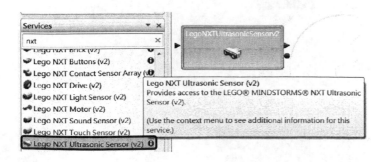

Step 2a—Modifying example 5.4, create a LEGO NXT Ultrasonic Sensor service component. This is the service component that corresponds to the actual ultrasonic sensor.

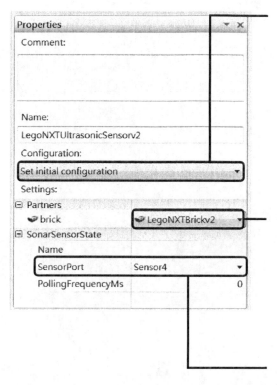

Step 2b—Set the initial configuration of the LEGO NXT Ultrasonic Sensor service component as **Set initial configuration**.

Step 2c—Set the connection component as the previously created LEGO NXT Brick service component.

Step 2d—Set the connection port as NXT Brick's Sensor4. (This will vary according to your own NXT package build.)

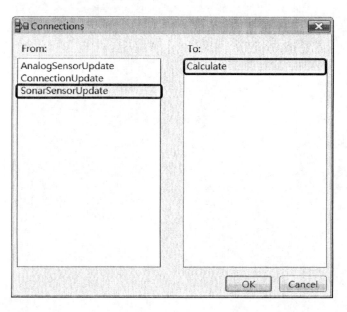

Step 3a—Connect the Notification of the **LEGO NXT Ultrasonic Sensor** service component to a **Calculate** activity component.

Step 3b—Set the connection from **SonarSensorUpdate** to **Calculate**; that is, output the sensed value of the ultrasonic sensor to the **Calculate** activity component.

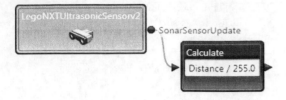

Step 3c—In the **Calculate** activity component, enter **Distance / 255.0**.

TIP

The LEGO NXT Ultrasonic Sensor service component's detection ranges from 0.0 (nearest) to 255.0 (farthest). A division by 255.0 normalizes the detection distance to between 0.0 and 1.0 and changes its significance such that the farthest state corresponds to 1.0 and the nearest state corresponds to 0.0. While the actual range of motor power is between –1.0 to 1.0, that is, it can rotate backward, we are only using the range from 0.0 to 1.0 so that it only moves forward.

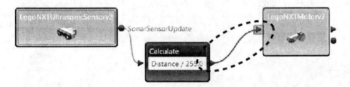

Step 4a—Connect the calculated value of the **Calculate** activity component to the **LEGO NXT Motor** service component.

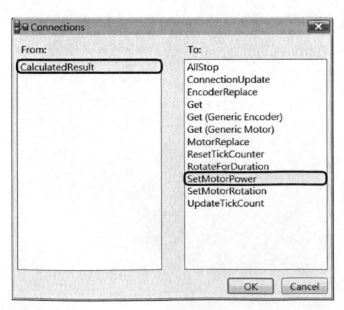

Step 4b—Set the connection from **CalculatedResult** to **SetMotorPower**, that is, using the calculated distance value to set the motor power.

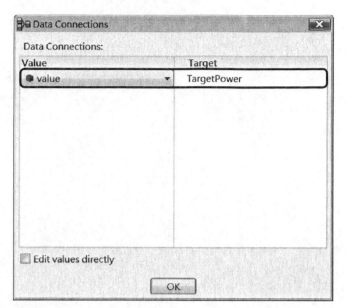

Step 4c—Set the corresponding variable relationship so that **value** corresponds to **TargetPower**, that is, the normalized distance value corresponds to motor power (the two ranges between 0.0 and 1.0).

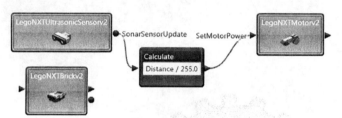

Step 5—The program is now complete.

After the program has started, the NXT Brick module will adjust the motor turning speed according to the distance detected by the ultrasonic sensor. When the robot is closer to an obstacle, the motor turning speed will decrease. This type of design is analogous to a cruising robot adjusting its speed according to its distance from an obstacle to avoid hitting the obstacle. This is a common design concept for robots.

5.4 EXERCISES

1. In example 5.4, if there was no active detection functionality, how would the behavior of the motor change?

2. Try to install two motors on the NXT module and assume that the left and right wheels of the robot control the motor. Now, rewrite examples 5.3 to 5.5 as follows:

    a. By rewriting example 5.3, can you make the robot avoid obstacles (e.g., set the robot to move backward when the touch sensor has been pressed)?

    b. By rewriting example 5.4, can you make the robot rotate in the dark to detect surrounding light sources?

    c. By rewriting example 5.5, can you make the robot turn to avoid an obstacle when it gets too close?

CHAPTER 6

Robot Motion Behavior

6.1 OVERVIEW

The purpose of this chapter is to introduce the motion behavior design of the LEGO robot, including the use of a game controller to control the robot or setting the robot to move autonomously according to the state of the environment. Given a group of robots that have been completely configured, you will learn the following from this chapter:

How to set MSRDS to remember the group's hardware configuration without the need to reset the program every time a program is to be written

How to control a real robot under the MVPL environment

How to integrate the sensors and the motors

How to let the robot roam autonomously

6.2 MANIFEST

In the example in the previous chapter, the NXT Brick module had either one or two supplementary hardware items (e.g., a light sensor coupled with a motor), and therefore setting corresponding service components was not a complicated task. However, when we have a complex robot, the task of setting service components becomes tedious and time-consuming. In fact, after the robot hardware design is complete, usually the configuration will not be changed, but if at each program redesign, the hardware associations within MVPL have to be rebuilt, the process becomes far more complex. For this reason, the manifest emerged as a preferred solution.

The design of the manifest serves to record specific hardware configuration relationships within MSRDS. The benefits it brings are as follows:

- **Elimination of the complexity of repeatedly building hardware configuration relationships**—Programmers can store a set of designed component relationships into the manifests, which are to be used for the different projects.

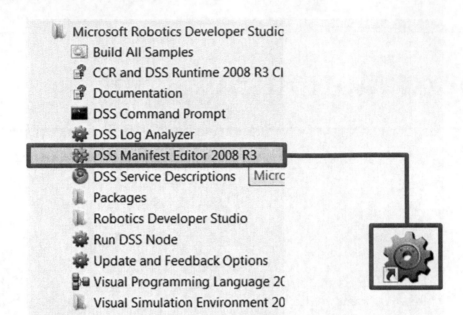

FIGURE 6.1   The DSS Manifest Editor within MSRDS.

- **Ability to address different hardware configurations with minor modifications—** If there are only a few resultant changes to the number of components and wiring methods when the design of the hardware components changes, only some minor adjustments are required to the manifest. As a result, programmers need not redesign hardware relationships from the start.

- **Ability to assist in the design of robot control system—**The same hardware information manifest can be passed on to different programmers or program teams for use, which assists in the development of robot functionalities.

There are several default manifests within MSRDS, stored in the **\samples\Config** folder under the installation directory,* with extension name **.manifest.xml**. These can be extracted by using the **import** command. To create the MSRDS hardware information manifest, the DSS Manifest Editor needs to be used. This tool can be found in the MSRDS program list (Figure 6.1). Since the DSS Manifest Editor and MVPL are located in the same program albeit with different start parameters, their runtime environments are very similar, as they are both divided into a similar command panel, service components, manifest, and properties (Figure 6.2). Apart from the lack of basic activity components, the DSS Manifest Editor has the same functionalities as MVPL.

We will begin by describing how to create the manifest based on the official recommended design for the LEGO robots (Figure 6.3), which will then be used in later examples.

---

* The default location of the installation directory is C:\Documents and Settings\{username}\Microsoft Robotics Dev Studio 2008 R3\ (for Microsoft Windows XP) or c:\Users\{username}\Microsoft Robotics Dev Studio 2008 R3\ (for Microsoft Vista and Windows 7), where {username} is the account you used to install MSRDS.

command panel   project panel

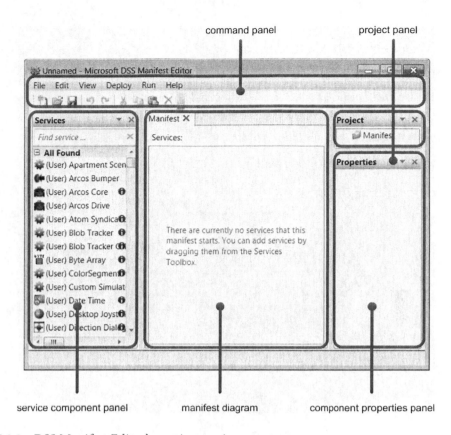

service component panel   manifest diagram   component properties panel

FIGURE 6.2   DSS Manifest Editor's runtime environment.

FIGURE 6.3   LEGO robot's official recommended design. (Please refer to the manual that comes with the package.)

EXAMPLE 6.1: CREATING THE ROBOT MANIFEST

Explanation: Set up the manifest for the LEGO robot official design (Figure 6.4).

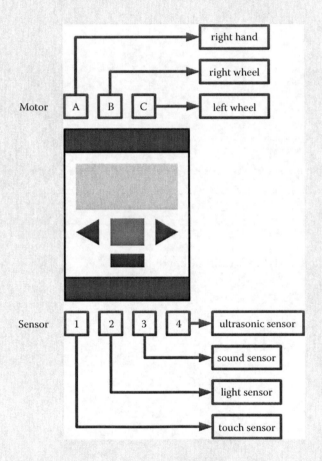

FIGURE 6.4 Connections between the NXT Brick module and other hardware components.

Skill: The ability to use the DSS Manifest Editor in MSRDS.

EXAMPLE 6.1 (continued): CREATING THE ROBOT MANIFEST

Completion diagram:

(Numbered balls represent the steps described below.)

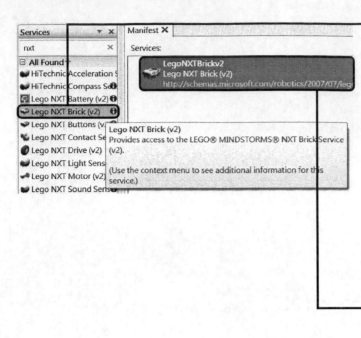

Step 1a—The interface and operation of the DSS Manifest Editor is similar to that for MVPL. First, select and create the **LEGO NXT Brick** service component from **Services**.

Step 1b—Select **LEGO NXT Brick** within **Manifest** area to edit its properties.

Step 1c—On the Properties page of the LEGO NXT Brick, select **Create Initial State**.

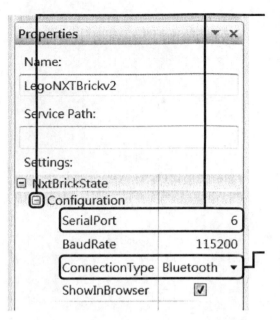

Step 1d—Click on the "+" icon to expand the configurations page and set the **SerialPort** to **6**. (This will vary according to the Bluetooth module connection settings on your own system.)

Step 1e—Tick the **ShowInBrowser** box to monitor the robot's situation during program execution in the browser.

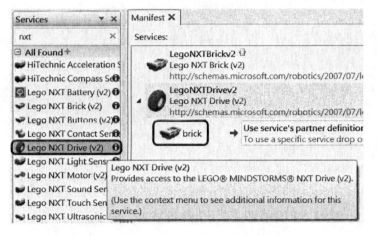

Step 2a—Next, create the **LEGO NXT Drive** service component from **Services**. This is similar to the **Generic Differential Drive** service component mentioned previously, apart from the fact that it is a drive component belonging to LEGO NXT and actually implements the **Generic Differential Drive** manifest. At this point, the connection for the **brick** component has not been set.

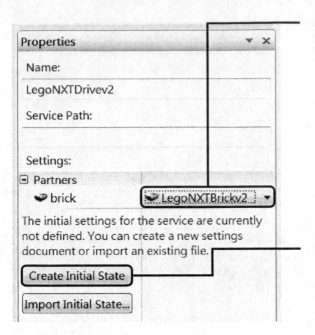

Step 2b—Click on the "+" icon to show the **Partners** page and set **brick** as the existing **LEGO NXT Brick** service component.

Step 2c—Select **Create Initial State** to create the initial state.

Step 2d—Click on the "+" icon to begin setting the **DriveState**.

Step 2e—Set the distance between the wheels, DistanceBetweenWheels, as **0.112** meters (or 11.2 cm).

Step 2f—Set the connection points with the NXT Brick module as **MotorC** (left wheel) and **MotorB** (right wheel) respectively, and set the wheel diameter as **0.055** meters (or 5.5 cm).

TIP

The values for distance between the wheels and wheel diameter depend on the different LEGO models you may build. If these parameters are not set correctly, the **RotateDegrees** and **DriveDistance** operations will not work correctly.

Step 2g—At this point, the connection settings between **LEGO NXT Drive** and **LEGO NXT Brick** service component have been set.

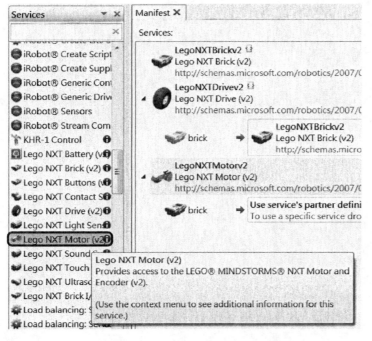

Step 3a—Create the **LEGO NXT Motor** service component from **Services**. This corresponds to the robot's right arm motor component.

Step 3b—Click on the "+" icon to show the **Partners** page and set **brick** as the existing **LEGO NXT Brick** service component.

Step 3c—Select **Create Initial State** to create the initial state.

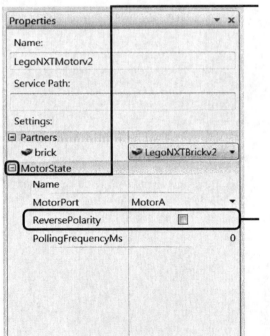

Step 3d—Click on the "+" icon to begin configuring the **MotorState**.

Step 3e—Set the connection point with the **NXT Brick** module as **MotorA**.

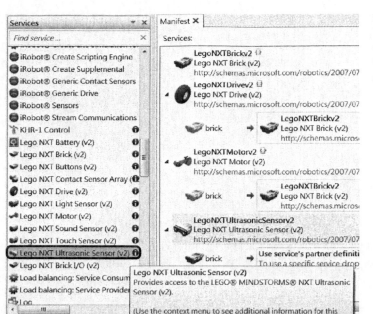

Step 4a—Select the **LEGO NXT Ultrasonic Sensor** service component in **Services** to create the ultrasonic sensor service component.

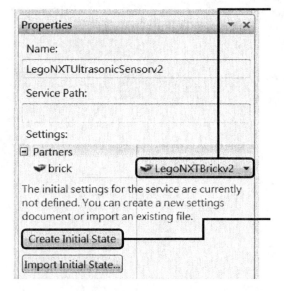

Step 4b—Click on the "+" icon to show the **Partners** page and set **brick** as the existing **LEGO NXT Brick** service component.

Step 4c—Select **Create Initial State** to create the initial state.

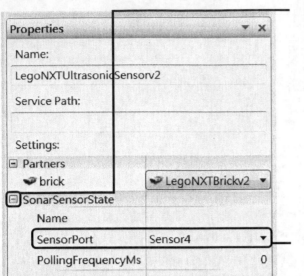

Step 4d—Click on the "+" icon to begin configuring the **SonarSensorState**.

Step 4e—Set the connection point with the **NXT Brick** module as **Sensor4**.

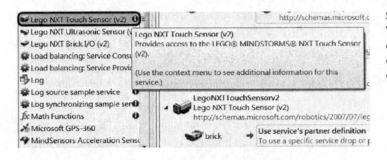

Step 5a—Select the **LEGO NXT Touch Sensor** service component from **Services** to create the touch sensor service component.

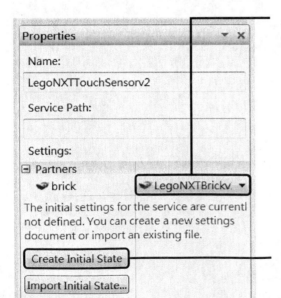

Step 5b—Click on the "+" icon to show the **Partners** page, and set **brick** as the existing **LEGO NXT Brick** service component.

Step 5c—Select **Create Initial State** to create the initial state.

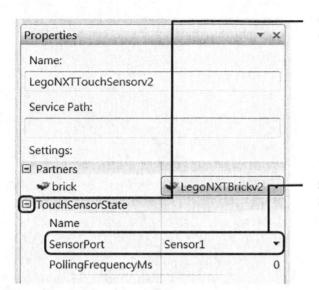

Step 5d—Click on the "+" icon to begin configuring the **TouchSensorState**.

Step 5e—Set the connection with **NXT Brick** module as **Sensor1**.

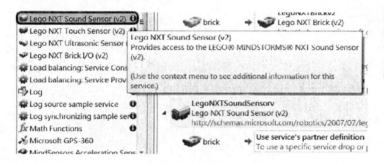

Step 6a—Select the **LEGO NXT Sound Sensor** service component from **Services** to create the sound sensor service component.

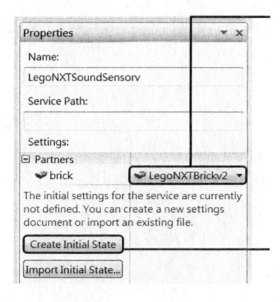

Step 6b—Click on the "+" icon to show the **Partners** page and set **brick** as the existing **LEGO NXT Brick** service component.

Step 6c—Select **Create Initial State** to create the initial state.

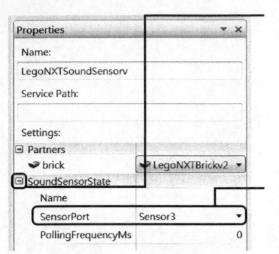

Step 6d—Click on the "+" icon to begin configuring the **SoundSensorState**.

Step 6e—Set the connection with the **NXT Brick** module as **Sensor3**.

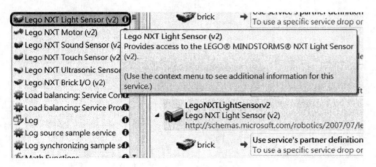

Step 7a—Select the **LEGO NXT Light Sensor** service component from **Services** to create the light sensor service component.

Step 7b—Click on the "+" icon to show the **Partners** page and set **brick** as the existing **LEGO NXT Brick** service component.

Step 7c—Click on **Create Initial State** to create the initial state.

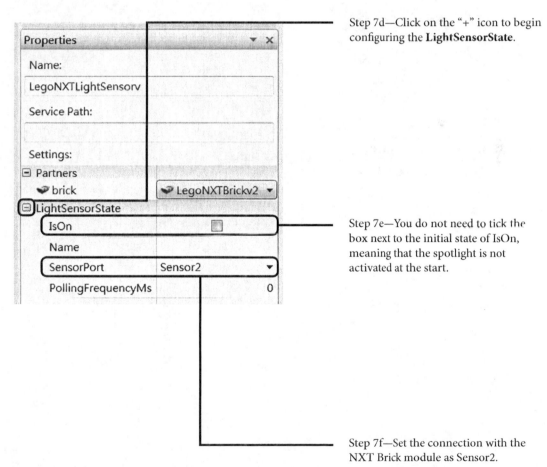

Step 7d—Click on the "+" icon to begin configuring the **LightSensorState**.

Step 7e—You do not need to tick the box next to the initial state of IsOn, meaning that the spotlight is not activated at the start.

Step 7f—Set the connection with the NXT Brick module as Sensor2.

Step 8—The manifest construction is now complete.

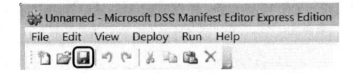

Step 9a—Click on the **Save** button to store the configured manifest.

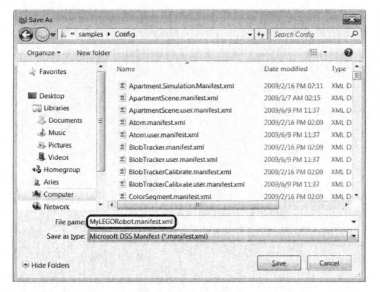

Step 9b—The save location can be set as the **\samples\Config** folder under the installation directory, storing the manifest with the existing manifests. This example will use the file name **MyLEGORobot. manifest.xml**

Step 10a—Click on **Run Manifest** or press the <F5> key to test the correctness of the newly created manifest.

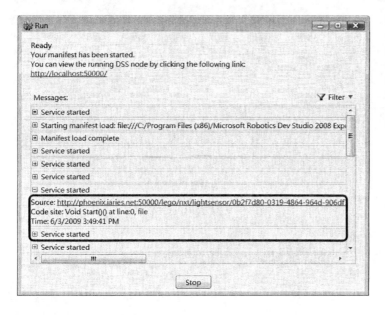

Step 10b—The background window of the DSS execution should display service components that have executed correctly, such as the light sensor. The corresponding hyperlinks can be used to monitor the status of the individual devices.

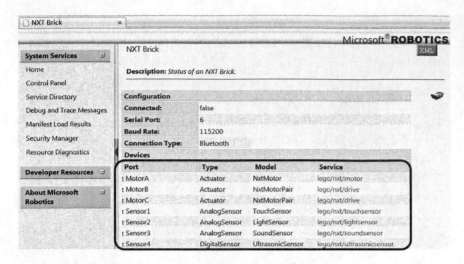

Step 10c—From the browser, we can also search and examine the state of every service component.

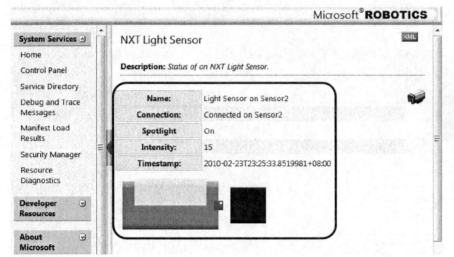

For example, the hyperlink for LightSensor links to a webpage showing the status of the **light sensor**.

6.3 AUTONOMOUS MOTION ROBOT PROGRAMMING

Chapters 4 and 5 have separately introduced the robot motion control programming under VSE and skills for individual hardware components. The combined knowledge of these techniques allows the control program to be written for the real robot. This chapter introduces two types of robot program design models: (1) the use of built-in control service components or commercial gaming controllers to control the robot; and (2) setting the robot to move autonomously by using values obtained from the sensors.

EXAMPLE 6.2: REMOTE ROBOT CONTROL

Explanation: Use the built-in control service component within MSRDS and the commercial gaming controller to remotely control the real robot. To do this, we will rewrite example 4.2.

Skill: The ability to use the manifest to control the real robot.

Completion diagram:

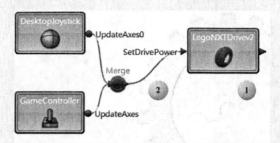

(Numbered balls represent the steps described below.)

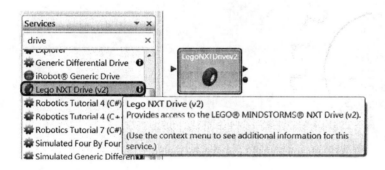

Step 1a—Create the **LEGO NXT Drive** service component.

Step 1b—In the **Properties** page, set the configuration to **Use a manifest**.

Step 1c—Choose Use existing or create new service for the manifest.

Step 1d—Select Import to import an existing manifest.

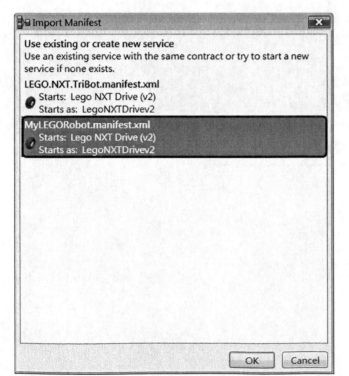

Step 1e—Select the **MyLEGORobot. manifest.xml** previously created.

Step 1f—The **Properties** setup is now complete.

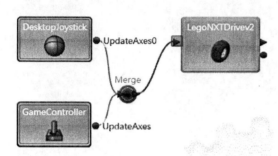

Step 2a—Delete the **Generic Differential Drive** service component from example 4.2 and direct the output of the **Merge** component to the **LEGO NXT Drive** service component.

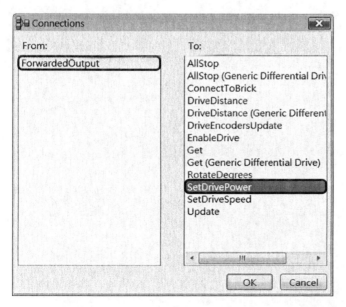

Step 2b—Set the connection type from **ForwardedOutput** to **SetDrivePower**.

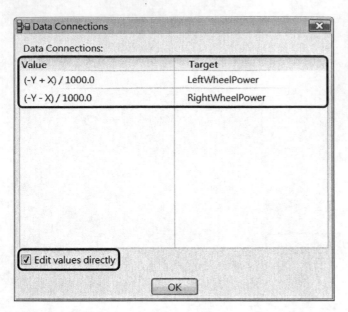

Step 2c—By clicking on the box next to **Edit values directly**, edit the connection value as **(-Y+X)/1000.0** for the left wheel power and **(-Y-X)/1000.0** for the right wheel power. The settings here are the same as that for examples 4.1 and 4.2.

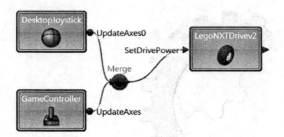

Step 3—At this point, the program design is complete.

After completing the above steps, the example is complete. Press the <F5> key to start the DSS service and use the Bluetooth module to connect to the real LEGO robot. During execution, a window identical to that of example 4.2 will appear and the user can simultaneously use the **Desktop Joystick** and gaming controller to control the real robot.

Robot Motion Behavior ■ 177

EXAMPLE 6.3: REMOTE CONTROL ROBOT (CONTROL OF MOTOR AND LIGHT SENSOR)

Explanation: Use a commercial gaming controller to control the real robot's motor and light sensor, by rewriting example 6.2.

Skill: The ability to use the manifest to control the real robot's motor and light sensor.

Completion diagram:

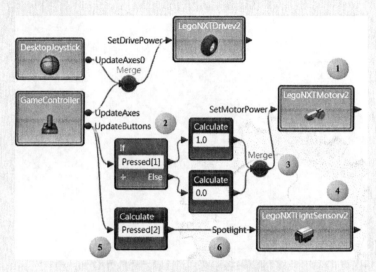

(Numbered balls represent the steps described below.)

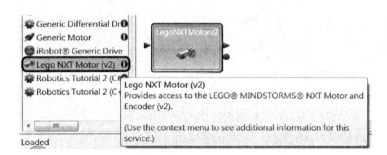

Step 1a—Expanding on example 6.2, create another **LEGO NXT Motor** service component.

Step 1b—Set the component's configuration as **Use a manifest** and set the manifest of **LEGO NXT Motor** in **MyLEGORobot.Manifest. xml**. This is our customized robot's real motor.

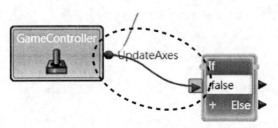

Step 2a—Connect the Notification of the **GameController** service component to a new **If** activity component.

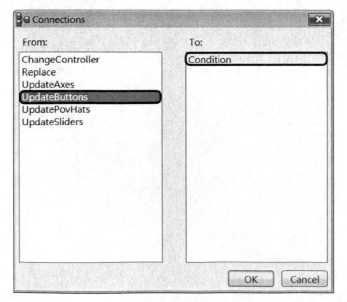

Step 2b—Set the connection from **UpdateButtons** to **Condition**.

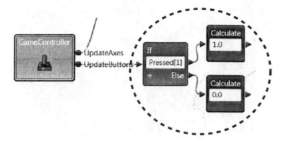

Step 2c—Set the condition of the **If** activity component as **Pressed[1]**. This corresponds to whether the gaming controller's second button is pressed and the output value is calculated as either **1.0** or **0.0**.

TIP

Pressed is a vector of the Boolean type, recording the "pressed" state of the gaming controller buttons. Conceptually similar to common programming languages such as C, the first corresponding value starts from **Pressed[0]** and so on. **Pressed[1]** therefore corresponds to the second button of the gaming controller.

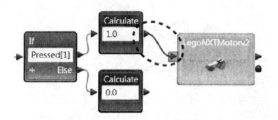

Step 3a—Connect the **Calculate** activity component to the **LEGO NXT Motor** service component.

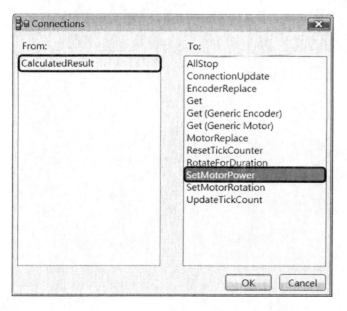

Step 3b—Set the connection from **CalculatedResult** to **SetMotorPower**, using the calculated result to set the motor power.

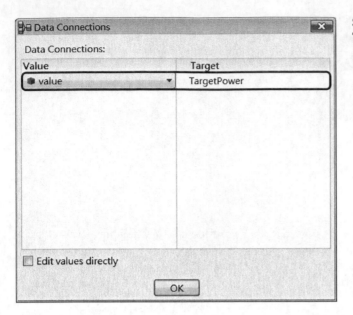

Step 3c—Set the **value** to the variable **TargetPower**.

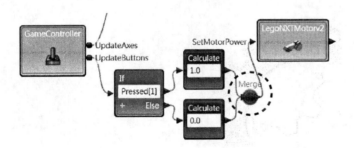

Step 3d—By using **Merge,** connect the two **Calculate** activity components to the **LEGO NXT Motor** service component, thereby completing the data flow for using the gaming controller's second button to control the real robot's right arm motor.

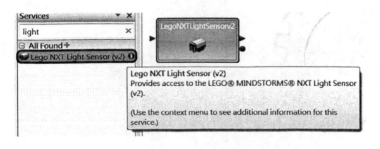

Step 4a—Create a **LEGO NXT Light Sensor** component.

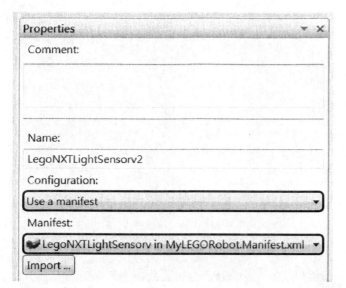

Step 4b—Set the component configuration as **Use a manifest** and set the manifest of **LEGO NXT Light Sensor** in **MyLEGORobot. Manifest. xml**. This is our customized robot's real light sensor.

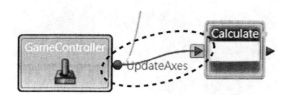

Step 5a—Connect the Notification of the **Game Controller** service component to a new **Calculate** activity component.

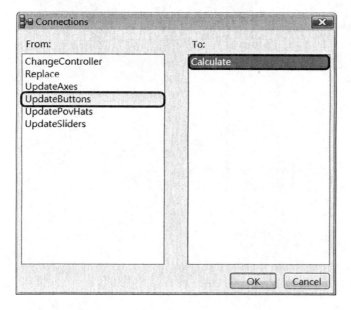

Step 5b—Set the connection from **UpdateButtons** to **Calculate**.

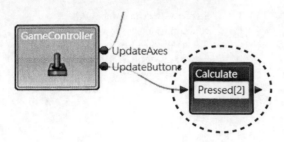

Step 5c—Set the **Calculate** activity component's calculation method as **Pressed[2]**. This corresponds to whether the gaming controller's third button is pressed and outputs a Boolean value.

Step 6a—Connect the calculated result of **Calculate** to the **LEGO NXT Light Sensor** component.

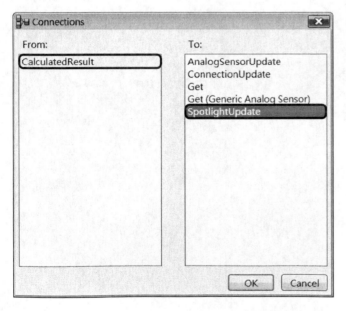

Step 6b—Set the connection from **CalculatedResult** to **SpotlightUpdate**, using the calculated Boolean value result to activate or deactivate the spotlight within the light sensor.

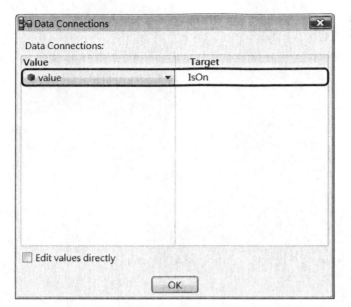

Step 6c—Set the **value** to the variable **IsOn**.

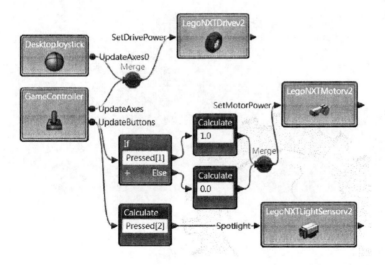

Step 7—The program design is now complete.

After completing the above steps, the example is complete. Press the <F5> key to start the DSS service and use the Bluetooth module to connect to the real LEGO robot. During execution, the gaming controller's second button controls the right arm motor, while the third button controls the light sensor's spotlight.

EXAMPLE 6.4: ULTRASONIC OBSTACLE AVOIDANCE ROBOT

Explanation: Use an ultrasonic sensor to control the robot's obstacle avoidance behavior.
Skill: The ability to combine the control capability of the ultrasonic sensor and the motor.
Completion diagram:

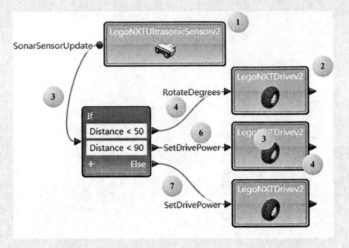

(Numbered balls represent the steps described below.)

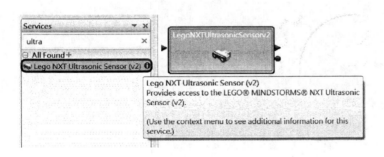

Step 1a—Create a **LEGO NXT Ultrasonic Sensor** service component.

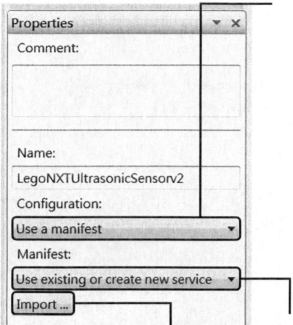

Step 1b—Set the component configuration as **Use a manifest**.

Step 1c—As this is the first service component in the new robot hardware configuration, the manifest has not been specified. Please select **Use existing or create new service** as the manifest and click on **Import** to import the manifest.

Step 1d—Choose the previously customized **MyLEGORobot. manifest. xml** manifest.

Step 2a—Create a **LEGO NXT Drive** service component.

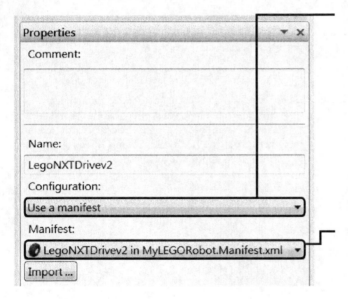

Step 2b—Set the component configuration as **Use a manifest**.

Step 2c—As the manifest was imported when setting the **LEGO NXT Ultrasonic Sensor** service component previously, we can directly choose **MyLEGORobot.Manifest.xml** as the manifest for **LEGO NXT Drive**.

Step 3a—Connect the Notification of the **LEGO NXT Ultrasonic Sensor** service component to an **If** activity component.

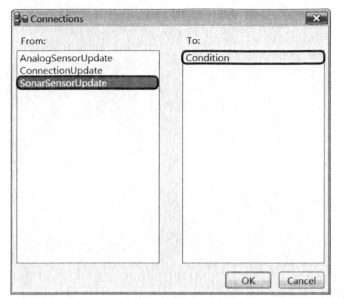

Step 3b—Set the connection from **SonacSensorUpdate** to **Condition**, directing the sensed result of the ultrasonic sensor to the decision field of the **If** activity component.

Step 4a—In the decision field of the **If** activity component, enter **Distance < 50**, where the **Distance** variable represents the distance value detected by the ultrasonic sensor. When this distance is less than **50** cm, the value is directed to the **LEGO NXT Drive** service component.

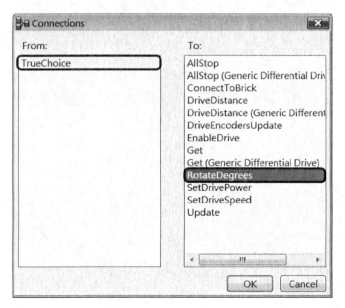

Step 4b—Set the connection from **TrueChoice** to **RotateDegrees** to control the robot's turning behavior.

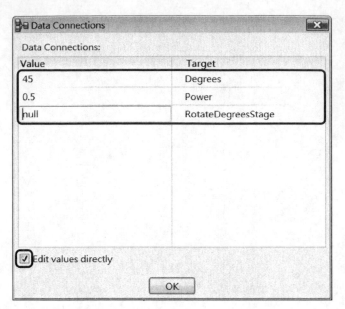

Step 4c—Click on the box next to **Edit values directly**. Set the value of **Degrees** as **45** degrees, the **Power** as **0.5**, and **RotateDegreesStage** as **null**. This sets the robot to turn 45 degrees counterclockwise with 0.5 units of power.

TIP

When the value of **RotateDegrees** is positive, the rotation will be counterclockwise; when it is negative, the rotation will be clockwise.

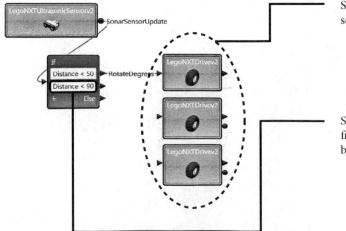

Step 5—Create three **LEGO NXT Drive** service components.

Step 6a—Add an additional decision field into the **If** activity component, being **Distance < 90**.

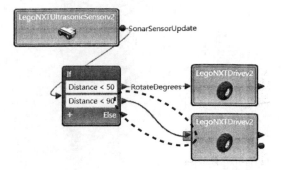

Step 6b—When the sensed value (**Distance**) of the ultrasonic sensor is greater than 50 cm but less than 90 cm, set the robot's corresponding motion behavior.

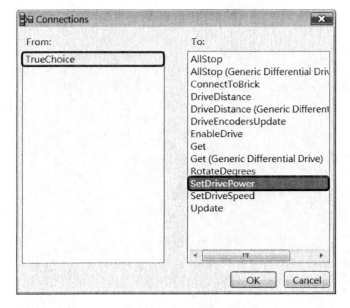

Step 6c—Set the connection from **TrueChoice** to **SetDrivePower**.

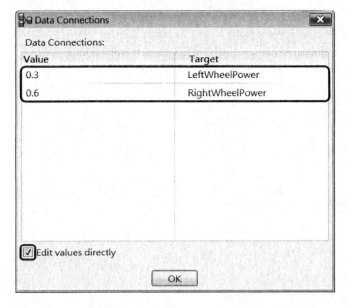

Step 6d—Click the box next to **Edit values directly**, setting **LeftWheelPower** as **0.3** and **RightWheelPower** as **0.6**. This causes the robot to turn left.

TIP

Although the previous configurations of both **RotateDegrees** and **SetDrivePower** appear similar, their behaviors are distinctly different. The function of **RotateDegrees** is to allow the robot to rotate at the same spot, whereas different left and right wheel power levels for **SetDrivePower** will cause the robot to move in circles (Figure 6.5).

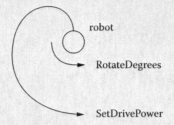

FIGURE 6.5 Difference in functionality between **RotateDegrees** and **SetDrivePower.**

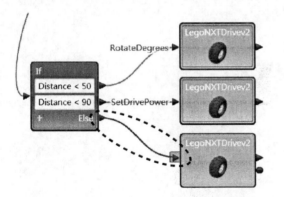

Step 7a—When the ultrasonic sensor's sensed value (**Distance**) is greater than 90 cm, set the robot's corresponding motion behavior.

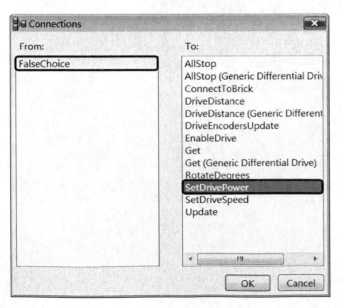

Step 7b—Set the connection from **FalseChoice** to **SetDrivePower**.

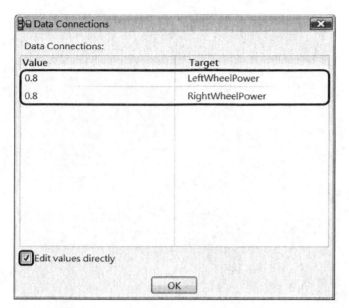

Step 7c—Click the box next to **Edit values directly**, setting both **LeftWheelPower** and **RightWheelPower** to **0.8**. This will make the robot move in a straight line.

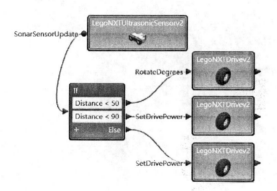

Step 8—The program design is now complete.

After completing the above steps, the example is complete. Press the <F5> key to start the DSS service and use the Bluetooth module to connect to the real LEGO robot. The robot will move autonomously in the following ways:

1. When the ultrasonic sensor detects an obstacle within 50 cm, the robot will turn 45 degrees counterclockwise while staying in the same position.

2. When an obstacle is detected within 90 cm, the robot will move to the left in a circular motion to avoid the obstacle.

3. When no obstacles are detected within 90 cm, the robot will continue moving forward.

EXAMPLE 6.5: LIGHT-SENSING-BASED ROBOT NAVIGATION

Explanation: Use light sensors to navigate a robot on a black rug with a diameter of 65 cm.
Skill: The ability to combine the control capabilities of the light sensor and motor.
Completion diagram:

1. Program overview

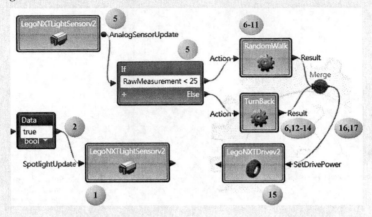

2. Random walk custom activity component

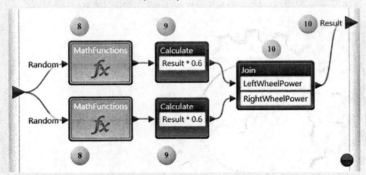

3. Turning back custom activity component

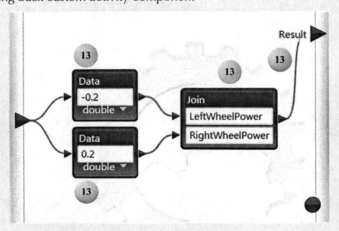

(Numbered balls represent the steps described below.)

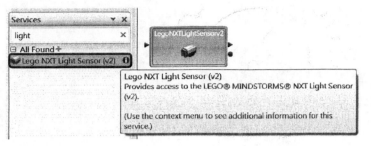

Step 1a—Create a **LEGO NXT Light Sensor** activity component.

Step 1b—Set the component configuration as **Use a manifest**.

Step 1c—As this is the first service component in the new robot hardware configuration, the manifest has not been specified. Please select **Use existing or create new service** as the manifest and click on **Import** to import the manifest.

Step 1d—Select the previously customized **MyLEGORobot.manifest.xml** manifest.

Step 2a—Create a **Data** activity component, using the Boolean value **true** to trigger the spotlight controlled by the **LEGO NXT Light Sensor** service component.

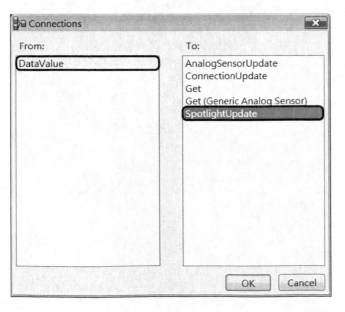

Step 2b—Set the connection from **DataValue** to **SpotlightUpdate**, using the input value to activate or deactivate the spotlight.

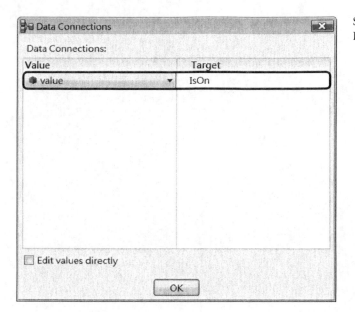

Step 2c—Set the **value** to the variable **IsOn**.

Step 3—As the first phase of this example is now complete, the program can be used to activate or deactivate the spotlight.

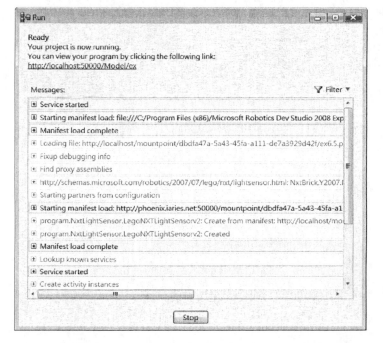

Step 4a—Press the <F5> key to start the DSS service.

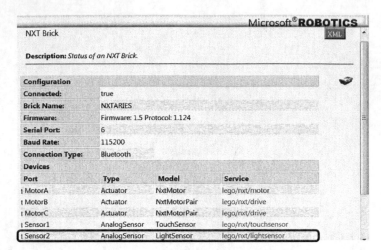

Step 4b—As the customized manifest has been set to start the browser, the monitoring screen is presented. In the browser, the connection of **LightSensor** can be selected to activate the light sensor's monitoring interface.

Step 4c—From the monitoring page of the light sensor, we can observe that currently the spotlight's state is **On** and that the sensed light intensity value is **54**. In this example, the robot is placed on a white surface and hence the light intensity value represents the corresponding sensed value for that environment's light source.

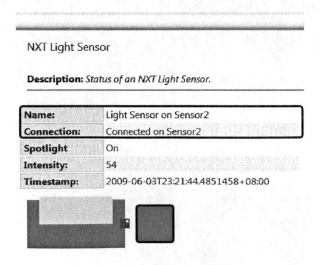

Step 4d—Move the robot to the black surface and refresh the sensor's monitoring page. The sensed light intensity has now decreased to **20**.

NXT Light Sensor

Description: *Status of an NXT Light Sensor.*

Name:	Light Sensor on Sensor2
Connection:	Connected on Sensor2
Spotlight	On
Intensity:	20
Timestamp:	2009-06-03T23:22:34.9221458+08:00

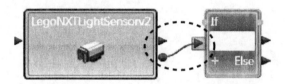

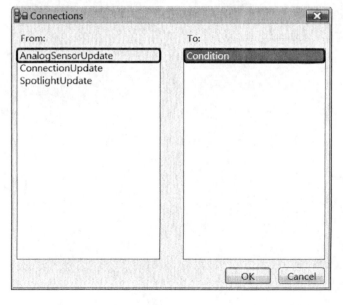

Step 5a—Going back to the program design, create another **LEGO NXT Light Sensor** service component duplicate and connect its Notification to an **If** activity component.

Connections

From:
AnalogSensorUpdate
ConnectionUpdate
SpotlightUpdate

To:
Condition

OK   Cancel

Step 5b—Set the connection from **AnalogSensorUpdate** to **Condition**, passing the sensed value to the **If** activity component for decision making.

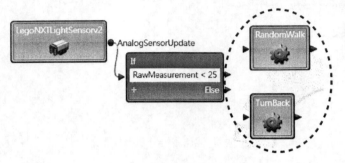

Step 5c—In the condition field of the **If** activity component, fill in **RawMeasurement < 25**. This will distinguish between the regions detected with sensed values lower than **25** (black rug) and values larger than **25** (areas outside the black rug).

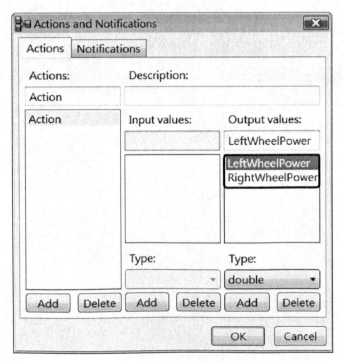

Step 6—Create two custom activity components, **RandomWalk** and **TurnBack** (the creation of custom activities was discussed in example 3.3), setting the functionalities to allow the robot to randomly walk and turn back.

Step 7a—First, modify the settings of the **RandomWalk** custom activity component.

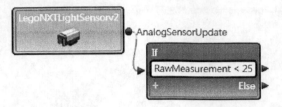

Step 7b—Set the output variables as two real numbers, being **LeftWheelPower** and **RightWheelPower**.

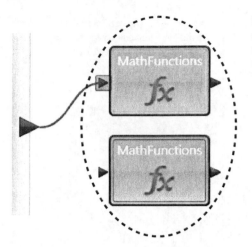

Step 8a—In the **RandomWalk** custom activity component, create two **Math Function** service components. When triggered, the **RandomWalk** custom activity component will start both of the **Math Function** service components.

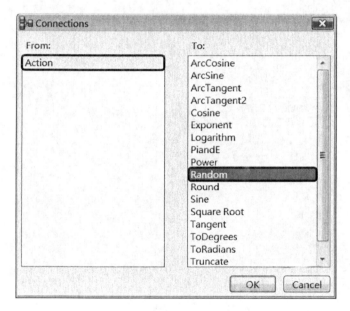

Step 8b—Set the connection such that when the **RandomWalk** custom activity component is triggered (**Action**), the **Random** function of **Math Function** is also triggered.

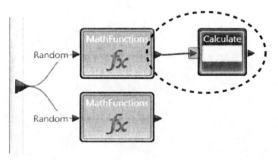

Step 9a—Connect the calculated result of **Math Function** to the **Calculate** activity component.

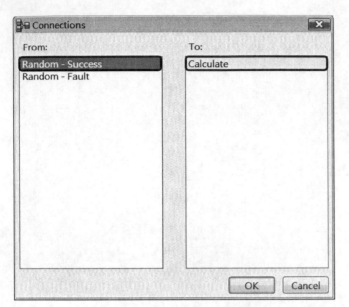

Step 9b—Set the connection from **Random - Success** to **Calculate**.

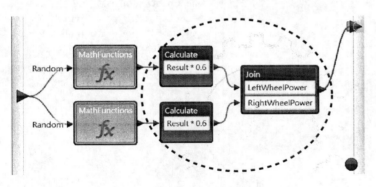

Step 10a—In **Calculate**, enter **Result*0.6**, doing the same in the other **Math Function** service component. Combine the results of the two **Calculate** activity components to one **Join** component and output the results accordingly.

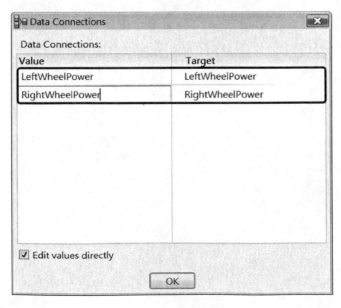

Step 10b—As the output variable names of the **RandomWalk** custom activity component are **LeftWheelPower** and **RightWheelPower,** which are the same as the **Target** variables, they will automatically match up.

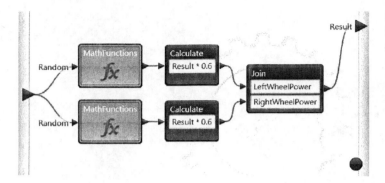

Step 11—At this stage, design of the **RandomWalk** custom activity component is complete.

Step 12a—Modify the settings of the **TurnBack** custom activity component.

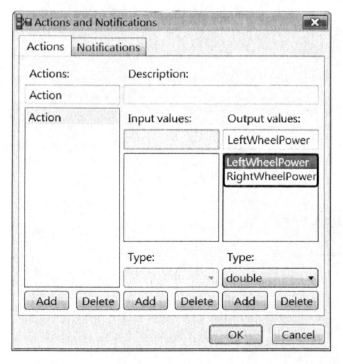

Step 12b—Set the output variables as two real numbers, being **LeftWheelPower** and **RightWheelPower**.

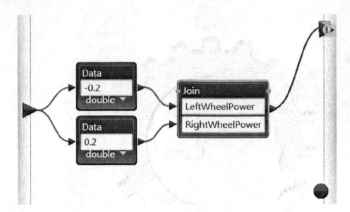

Step 13a—The settings for the **TurnBack** custom service component are simpler. The two **Data** activity components can be triggered directly, sending two real numbers –0.2 and 0.2 as output variable values.

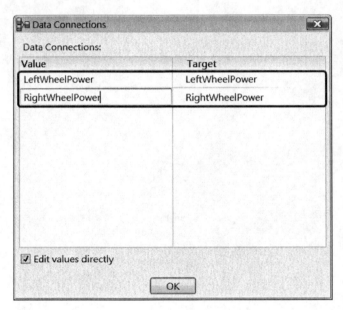

Step 13b—As the output variable names of the **TurnBack** custom activity component are **LeftWheelPower** and **RightWheelPower**, which are the same as the **Target** variables, they will automatically match up.

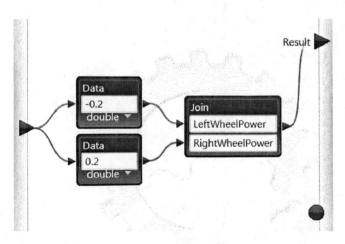

Step 14—At this stage, the design of the **TurnBack** custom activity component is complete.

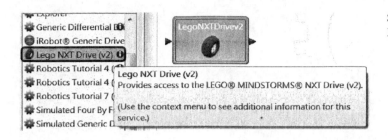

Step 15a—Create a **LEGO NXT Drive** service component.

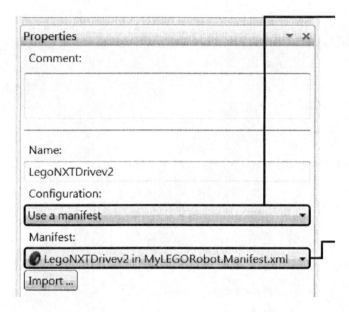

Step 15b—Set the configuration of the component as **Use a manifest**.

Step 15c—As the manifest was previously imported when setting the **LEGO NXT Light Sensor** service component, we can directly choose **MyLEGORobot.Manifest.xml** as the manifest for **LEGO NXT Drive.**

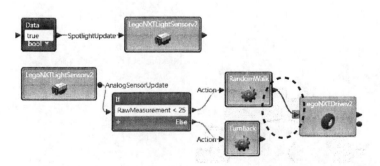

Step 16a—Connect the main program's **If** activity component to the **RandomWalk** and **TurnBack** custom activity components. Connect the outputs of those components to the **LEGO NXT Drive** service component.

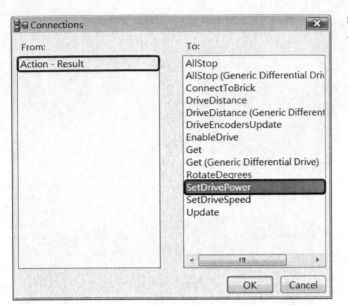

Step 16b—Set the connection from **Action - Result** to **SetDrivePower**.

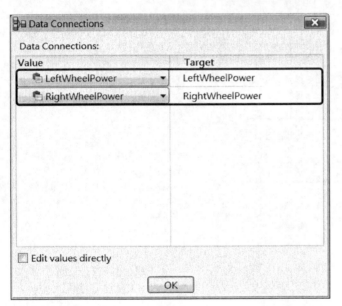

Step 16c—As the output variable names of the two custom activity components and **LEGO NXT Drive** service component are **LeftWheelPower** and **RightWheelPower**, which are the same as the **Target** variables, the variables are automatically matched.

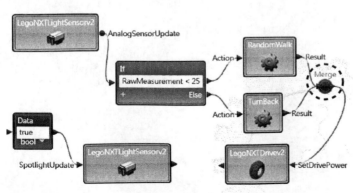

Step 17—Ensure that the **RandomWalk** and **TurnBack** activity components use the **Merge** technique to send data to the **LEGO NXT Drive** service component. The program design is now complete.

When the program is complete, press the <F5> key to start the DSS service and use the Bluetooth module to connect to the real LEGO robot. If the robot is initially placed on a black rug, the robot will navigate itself autonomously and will not go outside the rug area.

6.4 EXERCISES

1. Can you try to assemble your own LEGO education robot and construct its hardware manifest?

2. Expanding on example 6.4, can you give the robot different reaction behaviors? For example, can you make the robot move backward when it has sensed an obstacle 20 cm in front of it?

3. Expanding on example 6.5, can you set the robot to complete other actions during navigation? The following are some actions you may consider:

   a. When turning, emit a sound from the computer's speakers to report its own actions.

   b. Use the robot's ultrasonic sensor to detect obstacles during navigation.

CHAPTER 7

Controlling the Robot through Sounds

7.1 OVERVIEW

Service-type robots need to be capable of operating in changing environments and are primarily targeted for use by people who are not experts in technology; examples of service-type robots include homecare robots that primarily service the aged and entertainment robots that primarily service children. Therefore, it is important to determine ways in which robots can provide a more friendly interaction interface, such as controlling robots using sound signals. This chapter introduces the use of sound sensors and speech recognition capabilities. In this chapter, readers can develop their knowledge in the following areas:

- How to combine the sound sensor and motor
- How to use MSRDS to perform speech recognition
- How to use speech recognition to control a real robot

7.2 SOUND SENSOR

The sound sensor can measure the sound volume in the environment. We will illustrate its use through the following example.

EXAMPLE 7.1: SOUND-CONTROLLED SAMBA-DANCING ROBOT

Explanation: Design a samba-dancing robot, such that it can dance according to the volume of environment sounds.

Skill: The ability to combine the control capability of the sound sensor and motor.

Completion diagram:

1. Dancing robot's hardware design

2. MVPL program design

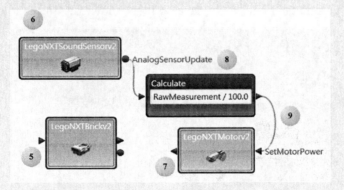

(Numbered balls represent the steps described below.)

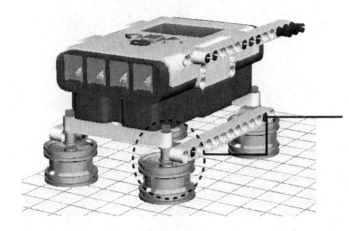

Step 1a—Use LEGO tires to assemble the robot's chassis.

Step 1b—Finely adjust the bar so that the height is at a suitable level.

Step 1c—Assemble the robot arm's fixed platform.

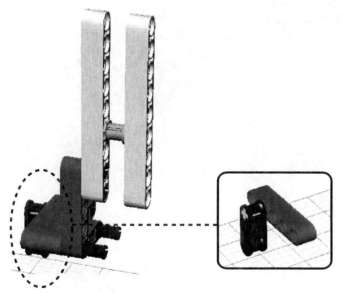

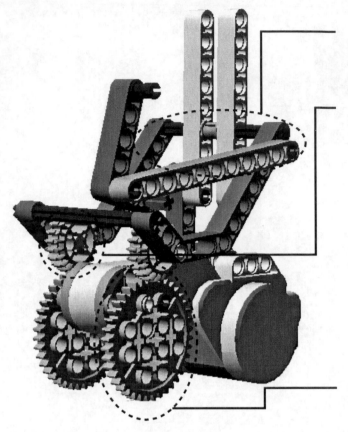

Step 2a—Assemble the activity bar. Inspect whether the two side gears may rub against the motor. If there is rubbing, please adjust the bar distance.

Step 2b—Assemble the cam-gear by plugging the yellow bolt into the center of the motor and the black bolt into the motor edge hole.

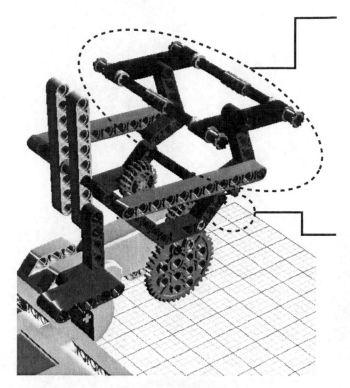

Step 2c—Install the item rack. After the installation is complete, the desired ornament or item may be placed and the rack can be adjusted accordingly.

Step 2d—Install an L-activity rack. Use a blue-colored bolt to tighten the linkage, ensuring that the blue bolt is not permanently fixed.

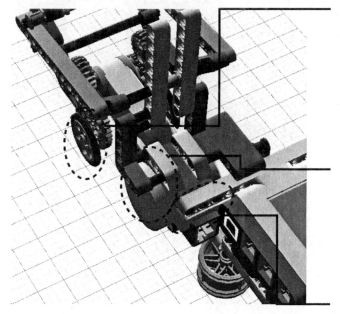

Step 3a—Install the axle on the side of the 40-tooth gear. The axles on both sides should be the same.

Step 3b—Use the L-shaped linkage to secure the frame.

Step 3c—Use three bolts to link the arm and NXT module body.

Step 4—The robot hardware design is now complete.

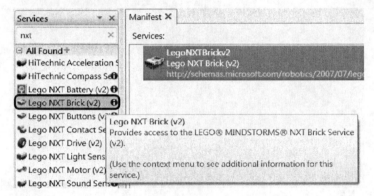

Step 5a—Create a **LEGO NXT Brick** service component.

Step 5b—At the properties page of the **LEGO NXT Brick**, click on **Create Initial State**.

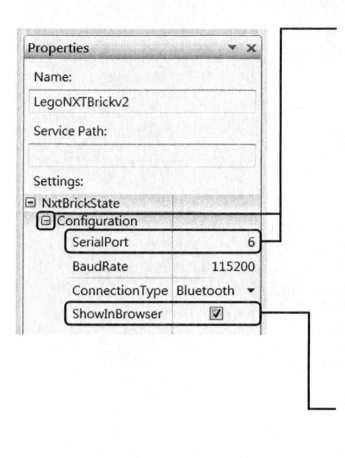

Step 5c—Click on the "+" icon to begin the configuration and set the **SerialPort** to **6**. (This will vary according to the actual Bluetooth module connection settings on your system.)

Step 5d—Tick **ShowInBrowser** to monitor the robot's situation on the browser during program execution.

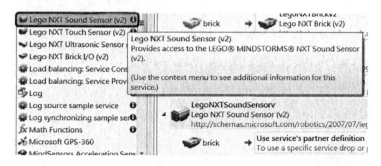

Step 6a—Select the **LEGO NXT Sound Sensor** service component in **Services** to create a sound sensor service component.

Step 6b—Select the "+" icon to show the **Partners** page and set the connecting brick as the existing **LEGO NXT Brick** service component.

Step 6c—Click on **Create Initial State** to create the initial state.

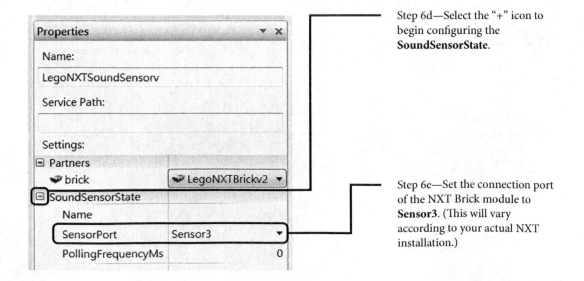

Step 6d—Select the "+" icon to begin configuring the **SoundSensorState**.

Step 6e—Set the connection port of the NXT Brick module to **Sensor3**. (This will vary according to your actual NXT installation.)

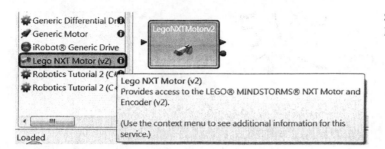

Step 7a—Create a **LEGO NXT Motor** service component.

Step 7b—Set the initial configuration of the **LEGO NXT Touch Sensor** service component to **Set initial configuration**.

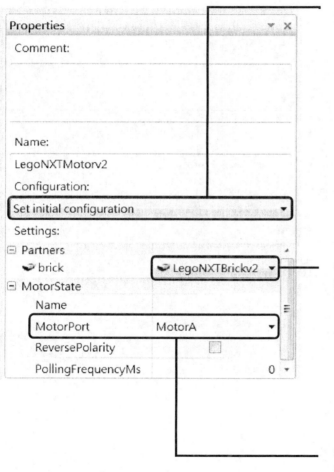

Step 7c—Set the connecting component as the previously created **LEGO NXT Brick** service component.

Step 7d—Set the connection port of the motor to **MotorA**. (This will vary according your actual NXT package build.)

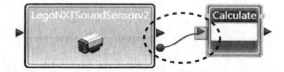

Step 8a—Connect the Notification of the **LEGO NXT Sound Sensor** service component to a **Calculate** activity component.

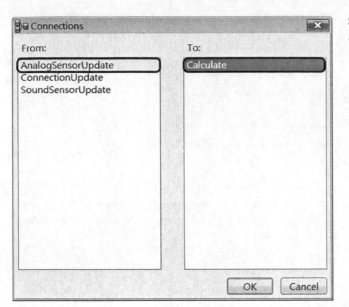

Step 8b—Set the connection from **AnalogSensorUpdate** to **Calculate**.

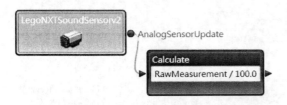

Step 8c—In the **Calculate** activity component, enter **RawMeasurement / 100.0**.

TIP

The value sensed by the sound sensor ranges from 0 (minimum) to 100 (maximum), and the power variable of the motor ranges from 0 to 1. Therefore, the **RawMeasurement** variable is divided by 100.0 within the **Calculate** activity component to match the motor power variable.

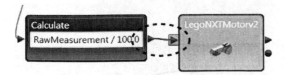

Step 9a—Connect the **Calculate** activity component to the **LEGO NXT Motor** service component.

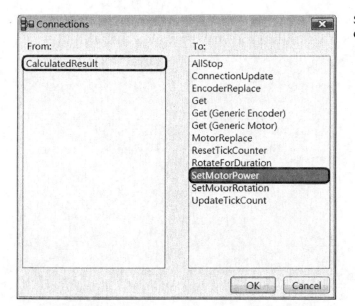

Step 9b—Set the connection from **CalculatedResult** to **SetMotorPower**.

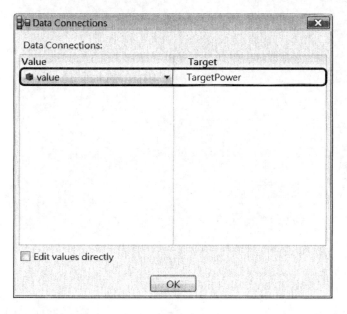

Step 9c—Set the corresponding variable relationship so that **value** corresponds to **TargetPower**.

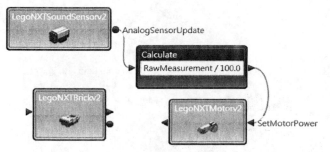

Step 10—The program design is now complete.

After completing the above steps, the example is complete. Press the <F5> key to start the DSS service and use the Bluetooth module to connect to the real LEGO robot. The robot will change its movement according to the volume of the sound in the surrounding environment.

7.3 VOICE CONTROL

Voice control is one of the interactive features of service robots. This chapter introduces the voice interaction function for LEGO robots under the MSRDS environment, including speech recognition and voice control.

EXAMPLE 7.2: SPEECH RECOGNITION

> **Explanation:** To test and demonstrate the speech recognition function in the MSRDS environment.
>
> **Skill:** The ability to operate the Speech Recognizer Graphical User Interface (GUI) and Speech Recognizer service components within MSRDS.
>
> **Completion diagram:**

(Numbered balls represent the steps described below.)

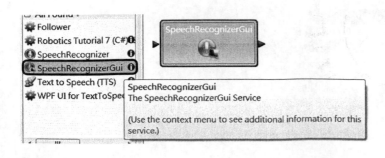

Step 1—Create a **SpeechRecognizerGui** service component. This is a GUI for setting parameters for speech recognition.

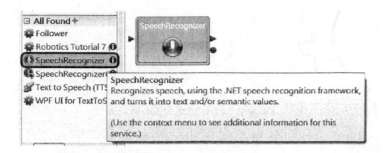

Step 2—Create a **SpeechRecognizer** service component. This is the speech recognition service component.

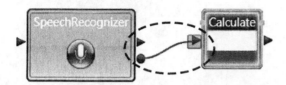

Step 3a—Connect the Notification of the **Speech Recognizer** service component to a **Calculate** activity component.

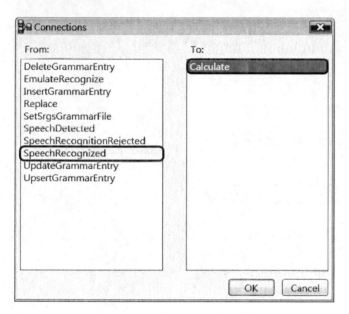

Step 3b—Set the connection from **SpeechRecognized** to **Calculate**.

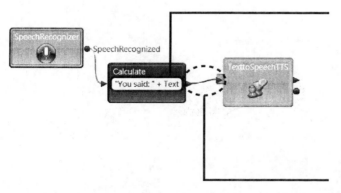

Step 3c—In the **Calculate** component, enter **"You said: " + Text**. Here, the **Text** variable is set as the result of **Speech Recognizer**.

Step 4a—Connect the result of **Calculate** to the **Text to Speech** service component.

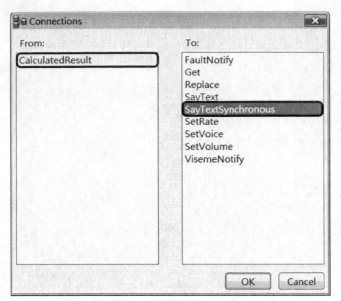

Step 4b—Set the connection from **CalculatedResult** to **SayTextSynchronous**, making the **Text To Speech** component read out the result of the **Calculate** activity component synchronously.

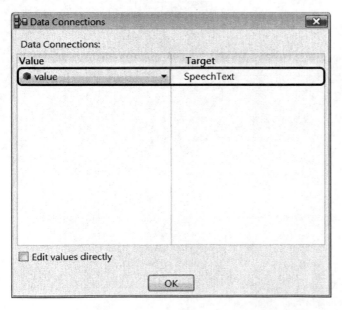

Step 4c—Set the variable relationship so that **value** corresponds to **SpeechText**. At this point, the program design is complete.

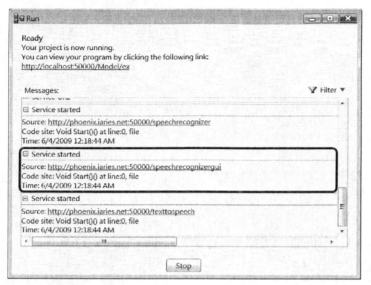

Step 5a—When the program is complete, press the <F5> to start the DSS service.

Step 5b—During execution, follow the interface to find records with **Service started**. Select the "+" icon here to look at the corresponding service component. We wish to find **http://localhost:50000/ speechrecognizergui**. Click on it to start the browser for monitoring the **SpeechRecognizerGui** service component.

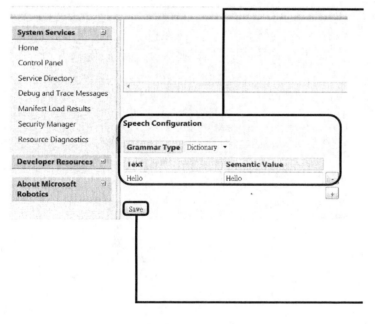

Step 5c—Look for **Speech Configuration** in the monitoring window for the **SpeechRecognizerGui** service component. Fill in **Hello** in both the **Text** field and the **Semantic Value** field.

Step 5d—Click **Save** to save the settings.

Step 6—Say the word "Hello" into the computer's microphone. The monitoring window will display the recognition result. If the recognition is successful, **You said: Hello** will be read out.

TIP

The **SpeechRecognizerGui** service does not need any connections and is only included in the diagram so that it will start automatically.

TIP

The voices that can be produced by the **Text to Speech** service component are limited by the operating system (in this case, Microsoft Window's system limitations). Currently, the built-in functionality only allows English voices to be produced, and additional purchase costs are required for voices in other languages. In addition, only Windows Vista and Windows 7 have a built-in speech recognition function. Although the **Speech Recognition** function is not automatically installed on Windows XP, it can still be installed from other products, such as Microsoft Office.

This example is used to test the computer's hardware and software (i.e., the microphone and operating system) for speech recognition functionality. If the speech recognition functionality of your computer performs normally, you can use your voice to control the real robot.

EXAMPLE 7.3: VOICE-CONTROLLED ROBOT

Explanation: Use the speech recognition service component to design a robot control system by rewriting example 7.2.

Skill: The ability to operate the Speech Recognizer GUI and Speech Recognizer service components within MSRDS.

Completion diagram:

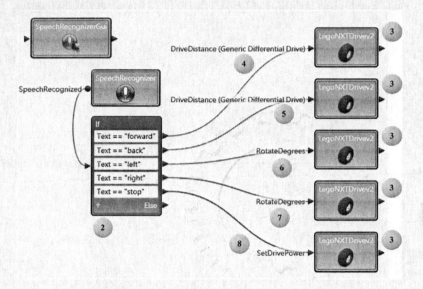

(Numbered balls represent the steps described below.)

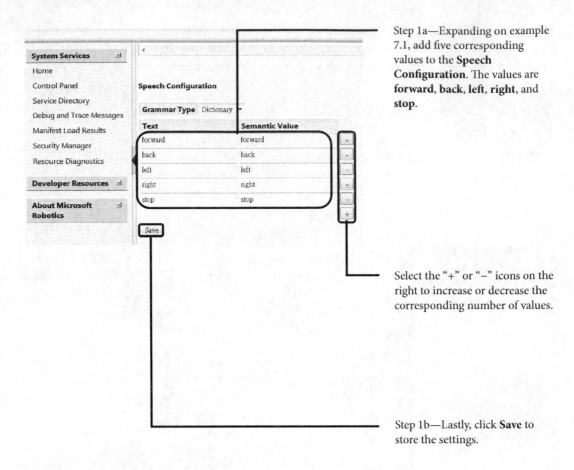

Step 1a—Expanding on example 7.1, add five corresponding values to the **Speech Configuration**. The values are **forward**, **back**, **left**, **right**, and **stop**.

Select the "+" or "–" icons on the right to increase or decrease the corresponding number of values.

Step 1b—Lastly, click **Save** to store the settings.

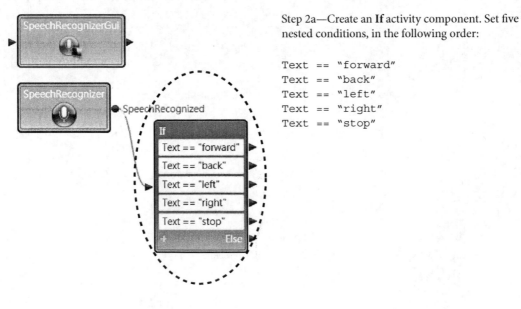

Step 2a—Create an **If** activity component. Set five nested conditions, in the following order:

```
Text == "forward"
Text == "back"
Text == "left"
Text == "right"
Text == "stop"
```

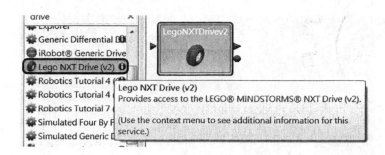

Step 3a—Create a **LEGO NXT Drive** service component.

Step 3b—Set the component's configuration as **Use a manifest**.

Step 3c—As this is the first service component for the new robot hardware component configuration, the manifest has not been specified. Please select **Use existing or create new service** as the manifest and click on **Import** to import the manifest.

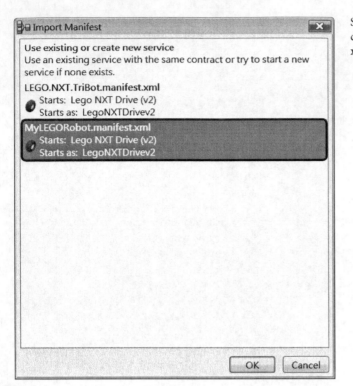

Step 3d—Select the previously customized **MyLEGORobot.manifest.xml** manifest.

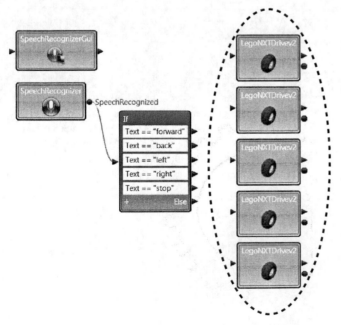

Step 3e—Duplicate the five **LEGO NXT Drive** service components.

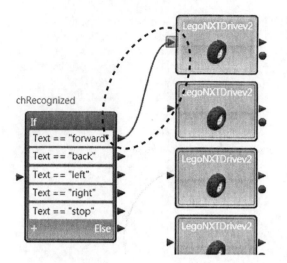

Step 4a—Connect **Text == "forward"** to the first
LEGO NXT Drive service component.

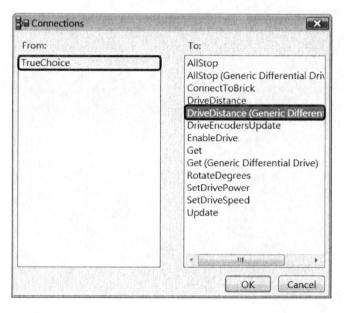

Step 4b—Set the connections from
TrueChoice to **DriveDistance
(Generic Differential Drive)** to control
the robot's drive distance.

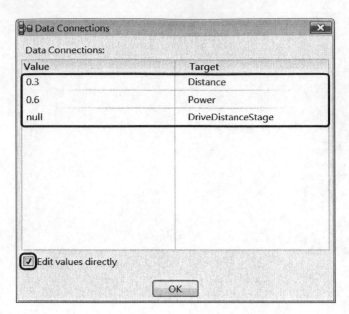

Step 4c—Click the box next to **Edit values directly**, setting the variable of **Distance** as **0.3**, **Power** as **0.6**, and **DriveDistanceStage** as **null**. This will instruct the robot to move a distance of 0.3 units with 0.6 units of power.

TIP

We chose the **DriveDistance** (**Generic Differential Drive**) service component rather than the **DriveDistance** service component because the former requires only the setting of three variables. Although it restricts the variety of movements, it is easier for beginners to learn.

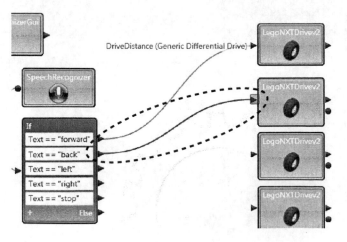

Step 5a—Connect **Text == "back"** to the second **LEGO NXT Drive** service component.

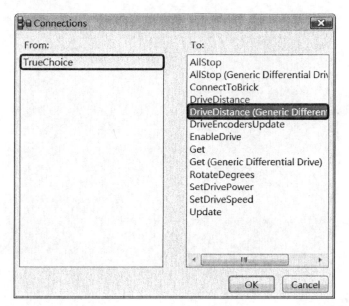

Step 5b—Set the connections from **TrueChoice** to **DriveDistance (Generic Differential Drive)** to control the robot's drive distance.

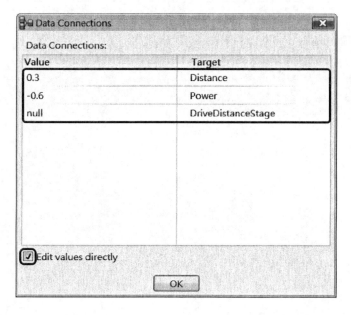

Step 5c—Click the box next to **Edit values directly**, setting the variable of **Distance** as **0.3**, **Power** as **–0.6**, and **DriveDistanceStage** as **null**. This will instruct the robot to move backward 0.3 units with 0.6 units of power.

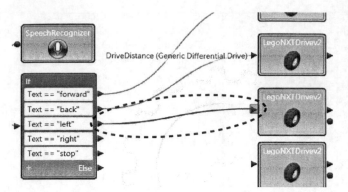

Step 6a—Connect **Text == "left"** to the third **LEGO NXT Drive** service component.

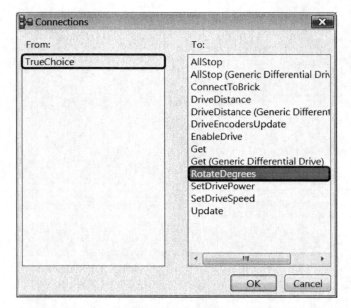

Step 6b—Set the connection from **TrueChoice** to **RotateDegrees** to control the robot's rotation behavior.

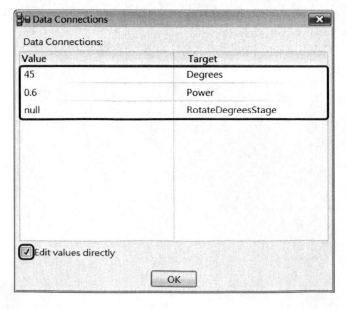

Step 6c—Tick the box next to **Edit values directly**, setting the variable of **Degrees** as **45**, **Power** as **0.6**, and **RotateDegreesStage** as **null**. This will instruct the robot to rotate 45 degrees counterclockwise using 0.6 units of power.

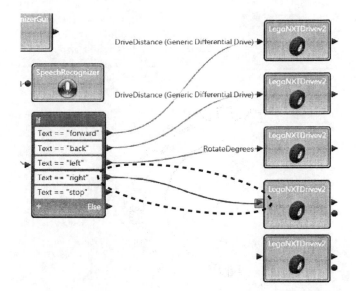

Step 7a—Connect **Text == "right"** to the fourth **LEGO NXT Drive** service component.

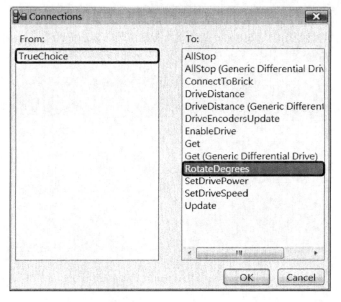

Step 7b—Set the connection from **TrueChoice** to **RotateDegrees**, to control the robot's rotation behavior.

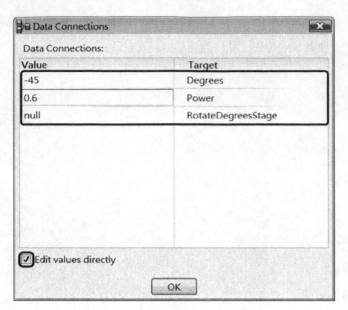

Step 7c—Click the box next to **Edit values directly**, setting the variable of **Degrees** as **–45**, **Power** as **0.6**, and **RotateDegreesStage** as **null**. This will instruct the robot to move clockwise 45 degrees with 0.6 units of power.

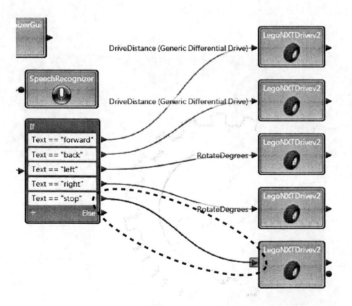

Step 8a—Connect **Text == "stop"** to the fifth **LEGO NXT Drive** service component.

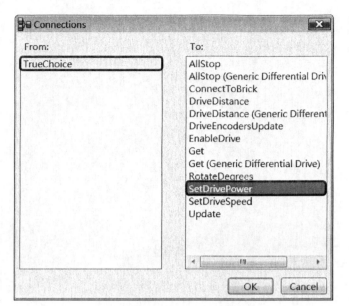

Step 8b—Set the connection from **TrueChoice** to **SetDrivePower**, to control the robot's wheel power.

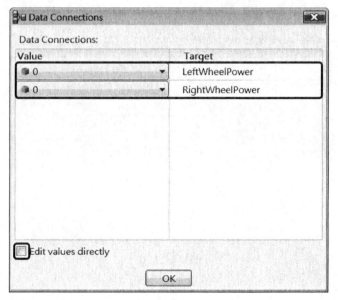

Step 8c—Tick the box next to **Edit values directly**, setting the variables for **LeftWheelPower** and **RightWheelPower** as **0**, instructing the robot to stop the rotation of both wheels.

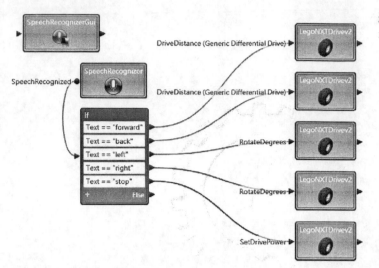

Step 9—The program design is now complete.

When the program is complete, press the <F5> key to start the DSS service and use the Bluetooth module to connect to the real LEGO robot. You can now say **forward**, **back**, **left**, **right**, and **stop** into your computer's microphone to control the robot's behavior!

7.4 EXERCISES

1. Expanding on example 7.1, can you set the robot to periodically perform actions other than those caused by the robot's reaction to the sound level? For example, can you make the robot vibrate at some frequency every 30 seconds?

2. Expanding on example 7.3, can you set the speech recognition service component to recognize more commands as well as coupling the commands with different behaviors such as a snake motion?

3. Apart from controlling the robot using sound, can you think of any other applications for which you can couple the speech recognition component with the robot?

CHAPTER 8

Robot Vision

8.1 OVERVIEW

Robot visual systems have been in development for many years. With the decrease in cost of visual recording components and the improved effectiveness of image processing techniques, the image recognition capability of robots is slowly being realized. Although LEGO robots are not packaged with visual recording components, video cameras can be integrated with the robot using MSRDS's integration functions to provide the robot with image recognition capability. While webcam products that use physical USB cables will restrict the robot's motion, an alternative is to use a wireless camera, which is slowly gaining in popularity. In this chapter, you can develop your knowledge in the following areas:

- How to integrate a video camera installed on the computer with MSRDS

- How to use the video camera to perform image recognition

- How to operate an image-tracking robot (lookout point)

8.2 ROBOT VISUAL RECOGNITION EXAMPLE

EXAMPLE 8.1: LOOKOUT POINT

Explanation: Combine a video camera with LEGO robots, allowing it to serve as a lookout point for tracking objects.

Skill: The ability to operate the webcam component and image recognition function within MSRDS.

Completion diagram:

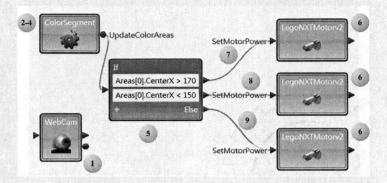

(Numbered balls represent the steps described below.)

Step 1a—Modify the LEGO robot's default assembly by installing a commercial video camera on the right arm motor. In this example, the camera connects to the computer using a USB cable. Please also ensure that the operating system drivers for your camera have been installed.

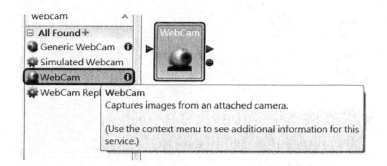

Step 1b—Create a **WebCam** service component.

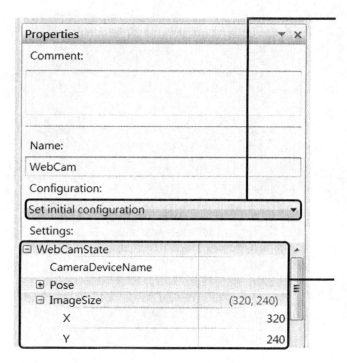

Step 1c—On the **WebCam** service component's **Properties** page, select the **Configuration** as **Set initial configuration**.

Step 1d—Expand on the **WebCamState** and **ImageSize** items. Set the image width and height for the video camera as **320** and **240** pixels respectively for X and Y.

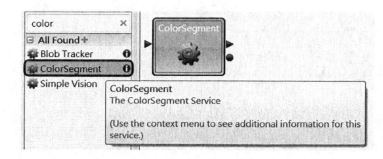

Step 2a—Create a **ColorSegment** service component. This component's function is to separate the image into color segments based on its color features.

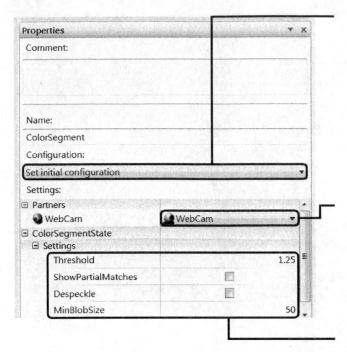

Step 2b—On the **Properties** page of the **ColorSegment** service component, set its **Configuration** to **Set initial configuration**.

Step 2c—Select to open the **Partners** item and set **WebCam** as the existing **WebCam** service component.

Step 2d—Select to open the Settings item and set the **Threshold** as 1.25, with the minimum blob size **MinBlobSize** as 50. The parameter **Threshold** is an allowable range for filtering color segments by the same color, and **MinBlobSize** is used for filtering out some small color segments.

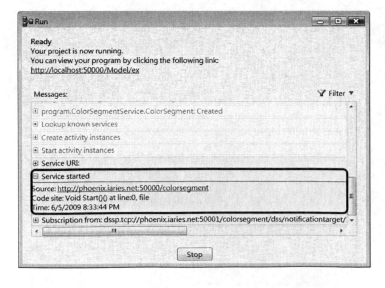

Step 3a—Press the <F5> key to start the DSS service. Search every line under **Service started** until you find the monitoring page for the **ColorSegment** service component. Click on the link for the monitoring page.

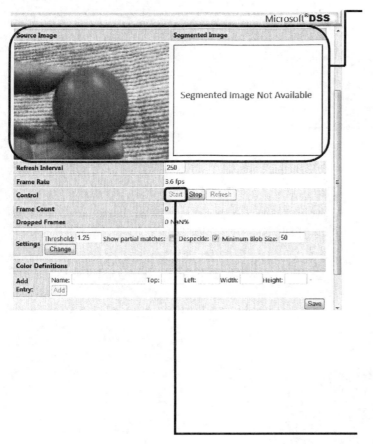

Step 3b—In the browser window, the **Source Image** column shows the camera's captured video. As an example, we can use the red ball from the LEGO robot package for image recognition. The segmented image has not been processed yet, and hence **Segmented Image Not Available** is displayed on the right.

Step 3c—Select the **Start** button from the **Control** column to let the video camera enter the detection mode.

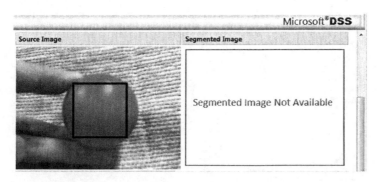

Step 3d—In the **Source Image** column, use the mouse to highlight and drag a rectangular box around a significant area of the red ball.

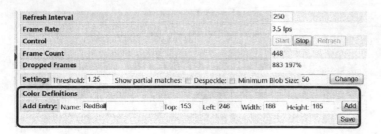

Step 3e—At the **Add Entry** position in the **Color Definitions** column, enter a custom name **RedBall** and click on the **Add** button.

TIP

Do not move the object before you click on the **Add** button, so that you can avoid any errors in the determined color.

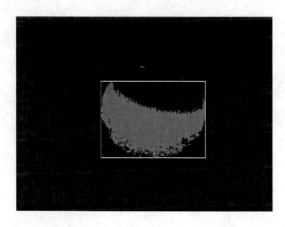

Step 3e—At this point, the **Segmented Image** column on the right will show the color segment that is recognized. The object may now be moved and the correctness of the tracking capability can be determined by whether the **Segmented Image** follows the moving image.

Refresh Interval	250
Frame Rate	2.7 fps
Control	Start Stop Refresh
Frame Count	1038
Dropped Frames	2009 193%

Settings Threshold: 1.25 Show partial matches: ☐ Despeckle: ☐ Minimum Blob Size: 50 [Change]

Color Definitions

RedBall Y, Cb, Cr: (124±52, 113±11, 189±31) R, G, B: (223, 82, 95) [Delete] [Expand Y]

Add Entry: Name: [ ] Top: [ ] Left: [ ] Width: [ ] Height: [ ] - [Add] [Save]

Step 3f—Make a note of the newly detected values for **RedBall**.

Step 4a—You can now stop the DSS service. Return to the **Properties** page of the **ColorSegment** service component and click on the "+" icon on the right of the **Colors** column to add a new color set definition.

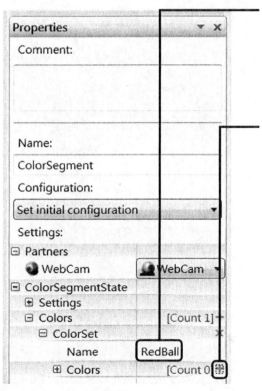

Step 4b—In the **ColorSet** column, set the **Name** as **RedBall**.

Step 4c—Click on the "+" icon to open the detailed definition column.

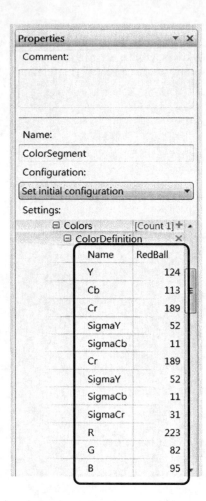

Step 4d—Enter the detection values you noted earlier in Step 3f in each field.

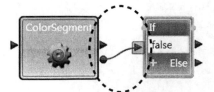

Step 5a—Return to the program edit area. Connect the Notification of the **ColorSegment** service component to an **If** activity component.

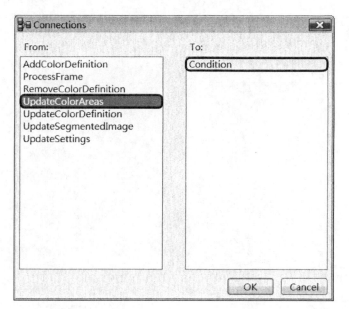

Step 5b—Set the connection from **UpdateColorAreas** to **Condition**, meaning that when the color segment region changes, the decision module of the **If** activity component is triggered.

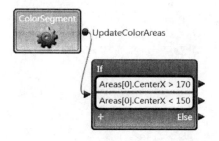

Step 5c—In the condition fields of the **If** activity component, enter **Areas[0].CenterX > 170** and **Areas[0].CenterX < 150**, representing three possible scenarios. The third scenario is the **Else** condition that indicates "150 < Areas[0].CenterX < 170".

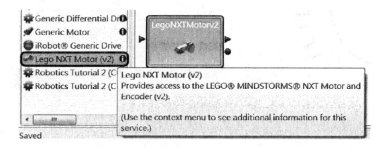

Step 6a—Create a **LEGO NXT Motor** service component.

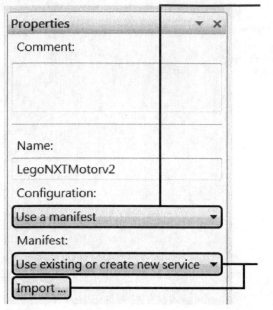

Step 6b—Set the component's configuration as **Use a manifest**.

Step 6c—As this is the first service component for the new robot hardware configuration, the manifest has not been specified. Please select **Use existing or create new service** as the manifest and click on **Import** to import the manifest.

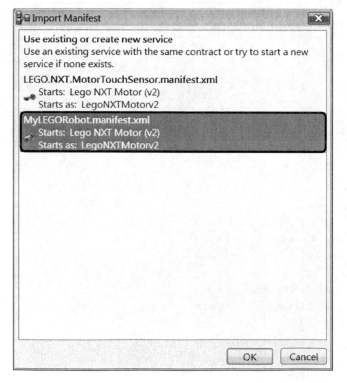

Step 6d—Choose the previously customized **MyLEGORobot.manifest.xml** manifest.

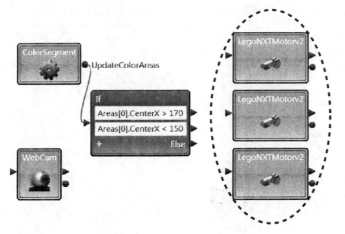

Step 6e—Create and duplicate three **LEGO NXT Motor** service components.

TIP

Areas is an array of color segment variables, sorted according to segment sizes. **Areas[0]** is the largest segment during recognition and, in this example, is most likely to reflect recognition of the red ball.

TIP

As the example uses an image width size of 320 pixels, **CenterX** is defined from 0 (furthest left) to 320 (furthest right). When the image center is greater than 170 pixels or less than 150 pixels, the red ball is considered to be outside the center, triggering the motor to drive to track the red ball.

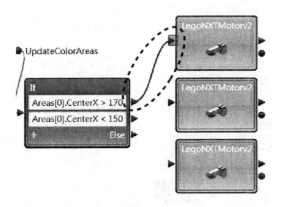

Step 7a—Connect the first situation of the **If** component, **Areas[0].CenterX > 170**, to the first **LEGO NXT Motor** service component.

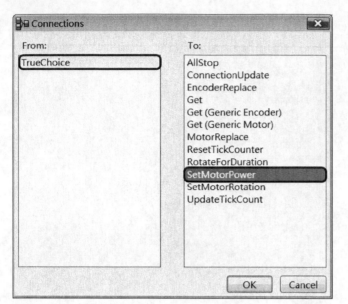

Step 7b—Set the connection from **TrueChoice** to **SetMotorPower**, referring to the motor power.

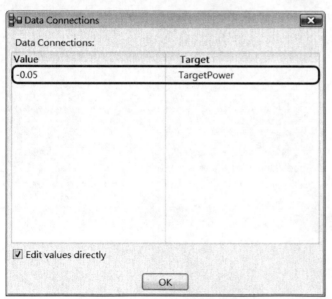

Step 7c—Set the corresponding variable relationship **TargetPower** as **−0.05**, which causes a counterclockwise rotation power of the motor at **0.05** units.

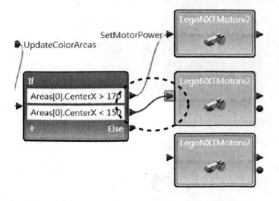

Step 8a—Connect the second situation of the **If** component, **Areas[0].CenterX < 150**, to the second **LEGO NXT Motor** service component.

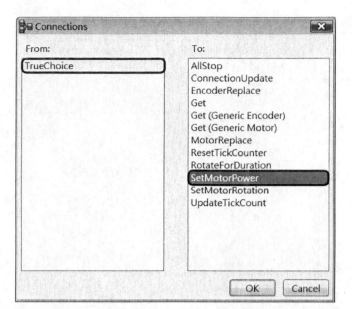

Step 8b—Set the connection from **TrueChoice** to **SetMotorPower**, referring to the motor power.

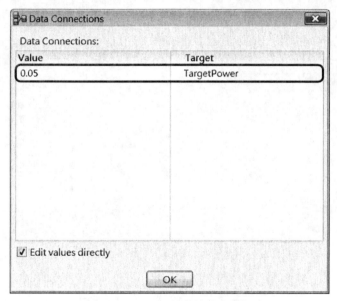

Step 8c—Set the corresponding variable relationship **TargetPower** as **0.05**, causing a clockwise rotation power of the motor at **0.05** units.

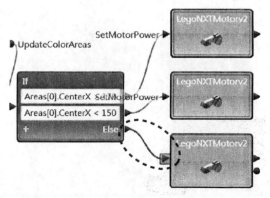

Step 9a—Connect the third situation of the **If** component to the third **LEGO NXT Motor** service component.

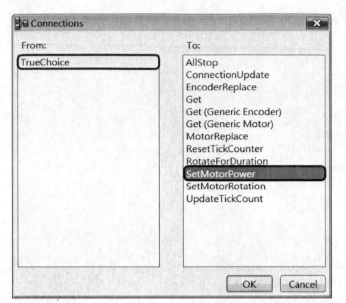

Step 9b—Set the connection from **TrueChoice** to **SetMotorPower**, referring to the motor power.

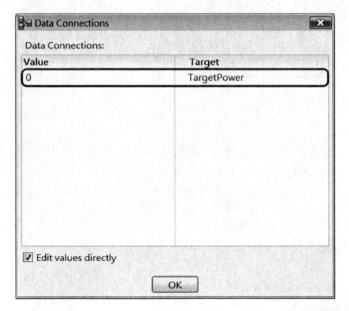

Step 9c—Set the corresponding variable relationship **TargetPower** as **0**, which stops the motor from turning.

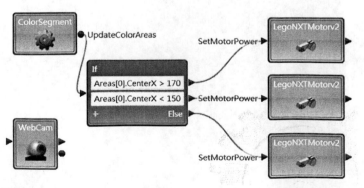

Step 10—The program design is now complete.

After completing the above steps, the example is complete. Press the <F5> key to start the DSS service and use the Bluetooth module to connect to the real LEGO robot. You can now move the red ball in front of the camera to cause the motor to track the red ball.

8.3 EXERCISES

1. Expanding on example 8.1, can you set the robot to react to its distance from the ball, in addition to tracking the red ball? (Hint: use the **ColorSegment** service component's **Areas[0].CenterY** parameter to program this.)

2. Can you design a control program to make the robot chase after the red ball?

3. Can you design a control program to enable the robot to distinguish between red and blue balls, and be able to push the chosen ball to a predetermined position?

CHAPTER 9

A Real Application— Sumo Robot Contest

9.1 OVERVIEW

Robotics study is an application-oriented field of study, and as such, it is interesting to apply robotics to real applications such as contests, which can often stimulate students' interest in robotics and help develop relevant skills. To this end, this chapter provides a real robotic application for educational purposes: the design of a sumo robot contest.

The robot platform used in this chapter is MSRDS, and all the hardware components are from the LEGO robots package. Students in this case example typically take 18 hours of class time to learn the knowledge and skills described in this book.

9.2 CONTEST DESIGN

The contest venue is made up of a black rug (diameter 65 cm) placed on a white surface and bordered by a square frame. The robot can determine its location information based on the surface color. The border dimensions are 300 cm by 300 cm and 15 cm high, ensuring that the robot will not be unfairly interrupted by external factors, such as interference from students (Figure 9.1).

The contest is divided into two parts, consisting of the practice exercise and the final competition.

9.2.1 Robot Design Review

Prior to the contest, the state of every robot will need to be tested. The test criteria include the following:

- Size restriction: The robot's body length cannot exceed 25 cm. This test method measures the robot from top to bottom and ensures that the whole body can be placed inside a 25 cm diameter circle. This restriction prevents any robot from having an advantage due to long arms.

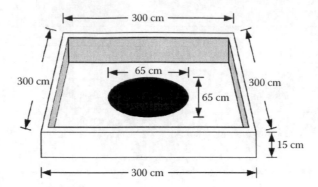

FIGURE 9.1   Sumo robot contest venue.

- Weight restriction: The robot's total weight is limited to the weight of all the components in one LEGO robot set. While students may use all the components in one LEGO robot set, they cannot use any LEGO parts that are additional to that one LEGO robot set, or any non-LEGO parts. This is to prevent any robot from winning the contest due to being relatively heavier, as the additional weight will mean it is much harder for the robot to be pushed out.

- Basic behavior: The robot must be able to perform tasks under the MVPL control such as move forward and backward; turn clockwise, counterclockwise, left, and right; and make reverse left turns and reverse right turns. This test is to ensure that each robot has the same basic operational abilities so that the contest can be focused on the behavior design and hardware construction of each autonomous robot.

- Robot design: Each group should design their own robot's motion behavior and contest strategy, such as preventing movement outside the rug area, defensive behavior on impact, and autonomous attack behavior. This is to ensure that each group's robot has its own individual characteristics, so that the winner will be determined based on the sumo robot with the strongest programming design.

9.2.2 Sumo Contest

After passing the first stage of robot testing, we enter the second stage of the contest. The contest uses a "two out of three wins" rule, in which the winner is promoted and enters the next round of the contest. Every round of the contest is limited to 2 minutes, and the resting interval between each round is also 2 minutes. The contestants can use the resting interval to tune the robot's hardware and software (with the requirement that the testing restrictions of the first stage are observed). A win can be determined in the following ways:

- When the winning robot pushes an opponent outside the contest area: The *area* refers to the rug, and the opponent is considered to have been pushed outside when the entire robot body is no longer on the rug.

- When the opponent is unable to continue in the contest: During the contest, a win will occur when the opponent is pushed down and is unable to continue operation or continue the contest.

- When the winning robot pushes down an opponent twice: The first successful push is when the opponent is overturned or its hardware component scatters during the fall. A subsequent successful push will result in a win for the round.

- Win determined by the cumulative number of pushes: When the round has ended and a win has not been determined, the number of pushes made by each robot is used to decide the win. A push is recorded when a robot pushes another robot for 2 seconds or more.

- If a result cannot be determined at the end of three rounds, the robot will enter an extension contest. The extension contest is won by the robot that makes the first push.

Every group of students will be given a set of contest rules as shown in Table 9.1.

9.3 ROBOT DESIGN

The real example explained in this section involves 13 students who were divided into six competing groups. The combination of students included five students (undergraduate and research level), two high school teachers, and six industry engineers (with job characteristics including machine design, integrated circuit design, screen/panel manufacturing, optical lens manufacturing, and so on). The members vary in their levels of understanding toward robotics, but each member has studied this textbook prior to participating in this contest. Each group's robot design is explained below.

The first group used an attack-oriented hardware design (Figure 9.2). It had the following features:

- The axle uses a direct drive approach. As the motor is used to turn the small gear, the large gear installed on the wheel increases the thrust and torque.

- A square repelling board is installed on the front, and the board is connected to the touch sensor. Upon impact with the opponent robot, it will thrust forward.

- Motion design: When the touch sensor has not been activated, the robot moves forward in a straight line and returns to the contest area when the light sensor detects that the robot is outside the area.

The second group used a defense-oriented hardware design (Figure 9.3). It had the following features:

- A low center of gravity.

- The sides of the robot include several slanted surface structures with an anticollision design, making it easier to overturn opponent robots that it comes into contact with.

- An ultrasonic sensor is installed on the front to detect the opponent robot. When the ultrasonic sensor discovers the enemy, the robot will sprint toward the opponent robot.

- Motion design: The robot first moves forward for some distance and then moves backward to the border. The light sensor installed on the back determines whether it has crossed the border.

TABLE 9.1   Robot Sumo Contest Rules

Part I: Structure and Behavior

Robot Design Rules

- Only use components in the LEGO robots package.
- Each robot can only use one LEGO robots package.
- The robot's body length cannot exceed 25 cm. The robot will be measured from top to bottom, ensuring that the whole body can be placed inside a 25 cm diameter circle.

Robot Design Acceptance (1.5 Hours) (Total 65%)

- The robot must be able to perform tasks such as move forward and backward; turn clockwise/counterclockwise, left, and right; and make reverse left turns and reverse right turns under the MVPL control. (15%)
- Description of robot structure and design concepts. (10%)
- Each group must design their own robot movement mode and contest strategy including attack (15%), searching for the opponent (15%), defense or avoidance (10%), and so on.

Part II: Sumo Contest (1 Hour) (Total 35%)

Contest time: _____ month _____ day _____ time _____ start, with the first 30 minutes used as preparation time.

Contest venue: _____

The contest venue is as shown below. The robots will be placed on a 65 cm diameter rug with their heads facing in the direction of the arrows.

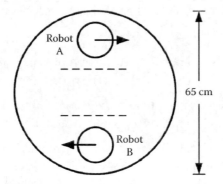

Contest Rules

- Every group is made up of two to three members, and an autonomous motion robot is to be used during the contest. It cannot be controlled using a controller or by voice recognition.
- The contest uses a two wins out of three rounds rule. Every round is 2 minutes and the rest interval between rounds is 2 minutes. Each group can adjust their robot hardware and software during the rest intervals but must observe the hardware design rules.
- To win, the robot must push the opponent outside the area (black mat) completely.
- You can also win by pushing down the opponent's robot, making it unable to move or continue participating in the contest.
- If you push down the opponent successfully two times, a win will also be awarded. A successful push is one where the components of the robot fall apart.
- During the contest, a point is recorded when the robot autonomously pushes the opponent's robot and the push lasts for 2 seconds or more.

TABLE 9.1 (continued)   Robot Sumo Contest Rules

- When the round is over and a win has not been decided, the accumulated points are used to determine the winner.
- If a failure to compete is due to software or hardware failure, the failure must be rectified within 3 minutes. Otherwise, the failure will be regarded as a defeat by withdrawal.
- If after three rounds a win is still not determined, an extension contest is given. The first push in this contest determines the winner.

Contest Schedule

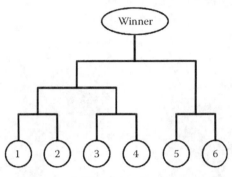

(Each group's ID is determined by a draw.)

FIGURE 9.2   The first group's sumo robot.

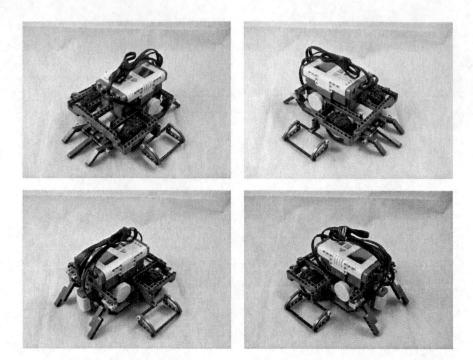

FIGURE 9.3   The second group's sumo robot.

The third group used an attack-oriented hardware design (Figure 9.4). It had the following features:

- It uses three motors and hence the body is heavier.

- The third motor drives the two white poles in front of the lower front area, allowing it to swing forward and backward.

- An ultrasonic sensor is installed on the front of the robot and used to detect if there is an enemy. If an enemy is found, the front white poles start swinging while pushing forward and backward. The body will also thrust forward in full power to try to push the enemy outside.

- Motion design: Turn 25 to 30 degrees every time, stopping for around 1 second after each turn to mitigate sonar reaction time problems.

The fourth group used an attack-oriented hardware design (Figure 9.5). It had the following features:

- It uses three motors and hence the body is heavier.

- The third motor drives the two gray poles at the front, allowing them to lift up.

- The ultrasonic sensor is installed at the front, while the touch sensor is installed at the back. This is to detect the presence of the enemy. When the enemy robot is

FIGURE 9.4   The third group's sumo robot.

FIGURE 9.5   The fourth group's sumo robot.

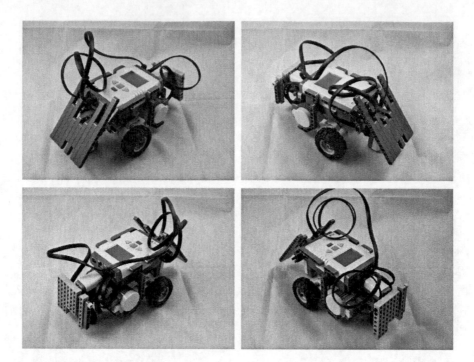

FIGURE 9.6   The fifth group's sumo robot.

detected at the front, it thrusts forward and lifts up the gray poles to disrupt the enemy's balance.

- Motion design: The robot will endlessly rotate in the same position.

The fifth group used a defense-oriented hardware design (Figure 9.6). It had the following features:

- The front of the robot has a blocking board for the purpose of defense as well as pushing.

- A touch sensor is installed at the back of the robot, with the pushing plate at the back used to drive the touch sensor to confirm whether the enemy is in contact behind the robot. Upon contact, it turns 180 degrees and then thrusts to push the enemy.

- Motion design: The robot will move forward and when the light sensor detects that it is out-of-bounds, the robot quickly reverses, rotates on the same spot, and then continues moving forward.

The sixth group was a design by the teaching assistance group (Figure 9.7), used to provide students with a reference design. It had the following features:

- A modification of the LEGO robot standard example.

- Slipping of the rear wheel is reduced when pushing the enemy.

FIGURE 9.7   The sixth group's sumo robot.

- Downward-facing slant structures are designed on the sides of the robot to guard against collisions by the enemy.

- Touch sensors are designed on the front to detect if the enemy is in contact. If contact is detected, the robot thrusts forward to push the enemy.

- A light sensor is installed at the front of the robot to prevent the robot from going out of bounds.

- Motion design: The robot has the random movement as described in example 6.5.

9.4 CONTEST RESULTS

Figure 9.8 shows the contest results, where the fourth group is the final champion. The key factors in their victory were the heavier weight of their robot's body and the use of a third motor to overturn opponents' actions and disrupt their balance.

Through such a contest, students obtained a real understanding of the design of robot behavior, including both software algorithms and hardware design. Apart from this, the contest brought a sense of stimulation that encouraged students' creativity and interest, which was beneficial to the learning of robotics.

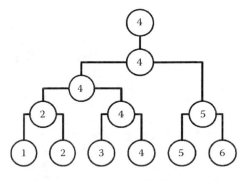

FIGURE 9.8   Contest process and results.

CHAPTER 10

Related Learning Resources

10.1 OVERVIEW

This chapter introduces learning resources related to the contents of this book which may be used as extended learning material. If you wish to obtain further knowledge after reading this book, the recommended websites in this chapter provide a wealth of knowledge for further reading.

10.2 RELATED COURSE WEBSITES

There are many robot integration courses in education institutions across the world. This book compiles a short list of courses available at well-known universities. You can download course materials and related programs from the given websites and conduct your own in-depth study.

Stanford University—Robot Programming Lab

- Institution: Stanford University

- Course code: CS 225B

- Course name: Robot Programming Lab

- Course website: http://cs225b.stanford.edu/

- Software used: Stanford University custom-designed robot software platform Player/Stage (Figure 10.1).

- Hardware used: wheel-type robot (Figure 10.2)

- Course aim: to teach intelligent robot control and perception, allowing students to learn how to adapt robots to complex situations in the real world, overcoming uncertainty to accomplish set tasks.

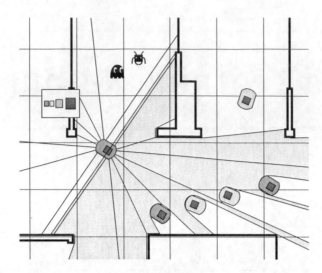

FIGURE 10.1   Stanford University custom designed robot software platform Player/Stage.

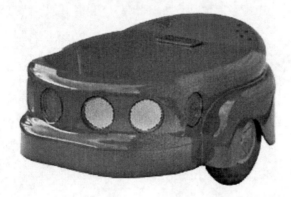

FIGURE 10.2   Wheel-type robot.

Massachusetts Institute of Technology—Autonomous Robot Design Contest

- Institution: Massachusetts Institute of Technology (MIT)

- Course code: 6.270

- Course name: Autonomous Robot Design Contest

- Course website: http://web.mit.edu/6.270/

- Software used: use of the C programming language to develop the software

- Hardware used: Happyboard (Figure 10.3)

- Course aim: to teach related robotic software/hardware; to train students to integrate software and hardware for learning autonomous robot information and movement control; and lastly, to design a complete robot.

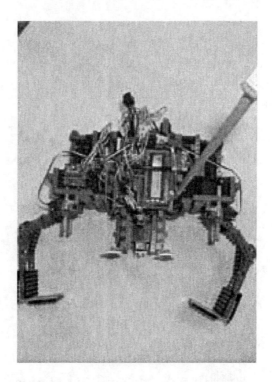

FIGURE 10.3   A MIT designed, homemade robot with Happyboard.

Carnegie Mellon University—Creating Intelligent Robots

- Institution: Carnegie Mellon University (CMU)

- Course code: 15-491

- Course name: Creating Intelligent Robots

- Course website: http://www.andrew.cmu.edu/course/15-491/

- Software used: Microsoft robot development platform MSRDS

- Hardware used: well-known robots such as iRobot Create and AIBO (Figure 10.4)

- Course aim: to teach basic movement control and perception sensors in intelligent robot's software design, focusing on topics such as robot self-learning and intelligence as well as problems such as robot visuals, path planning, and robot collaboration.

National Taiwan University—Robotics in Construction Automation

- Institution: National Taiwan University (NTU)

- Course code: 512EU8720

- Course name: Robotics in Construction Automation

- Course website: http://robot.caece.net/

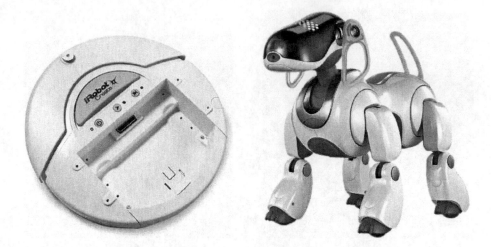

FIGURE 10.4   iRobot Create robot (left) and AIBO dog robot (right).

- Software used: Microsoft robot development platform MSRDS

- Hardware used: LEGO robots

- Course aim: This course is offered by the Civil Engineering division of National Taiwan University. It introduces the construction of automated robots and allows students to apply robots in the domain of civil engineering, such as automated road testing or hoisting projects.

10.3 OTHER WEBSITE RESOURCES

The Internet has several communities or forums related to MSRDS or LEGO robots. We list the following websites as recommended references. You can use these websites to find the required resources or even find people to exchange knowledge within these communities.

Microsoft Robotics Developer Studio robot development platform (http://msdn.microsoft.com/en-us/robotics/)—Microsoft robot development platform's main website provides MSRDS 2008 R3 Edition for installation download and contains related MSRDS teaching websites. You can follow the teaching websites to study MSRDS's design and development in more detail or become involved in the forums for further discussion.

Microsoft Robotics Developer Studio teaching files (http://msdn.microsoft.com/en-us/library/bb881626.aspx)—This website contains MSRDS's teaching files on MSDN. The content ranges from fundamental CCR and DSS teaching material to real applications such as how to set up and use service components, VSE, MVPL, and hardware connections. You can use this website to learn more about MSRDS technologies.

Microsoft Robotics Developer Studio development team Wiki page (http://channel9.msdn.com/wiki/msroboticsstudio/)—The MSRDS design team's Wiki page. From this website, the reader can see information related to the MSRDS team's internal developments. This website is also regularly updated to show new technologies, teaching files, and important technology documents related to MSRDS.

Microsoft Robotics Developer Studio forum (http://social.msdn.microsoft.com/forums/en-US/category/robotics/)—This is the Microsoft robot development platform forum, which is a place to discuss related software or hardware technologies. If you have any questions while using MSRDS, they can be posted to the development team at this forum, and the team will attempt to solve your problems. This forum also lists problems and suggestions contributed by MSRDS users around the world.

Young Joon Kim's Community (http://www.helloapps.net/)—This is the teaching website created by an ex-MSRDS team member, Young Joon Kim. It uses simple programming code and MVPL for teaching. The reader can learn CCR, DSS, and VSE development skills from this website.

LEGO Robots official page (http://mindstorms.lego.com/)—LEGO Robots' official page, which includes a basic introduction, development software download, hardware installation, and related product information. You can find the newest information about related LEGO robots from this website.

LEGO Technic Torano Maki (http://www.isogawastudio.co.jp/legostudio/toranomaki/en/index.html)—The website presents the LEGO hardware assembly introduction book, LEGO Technic, written by Isogawa Yoshihito, whose content provides a complete introduction and explanation of the LEGO hardware structure. A free version is available for download, and if the reader is willing, US$10 may be contributed to the author. From the book, you can obtain inspiration for LEGO robot structure designs and create robots that can perform more varied tasks.

10.4 RELATED BOOKS

For readers who are native Chinese speakers, the original edition of this book is shown in Figure 10.5. When you have finished learning the content of this book, you can look up the book titled *Programming Microsoft Robotics Studio*, which was written by the MSRDS development team, to obtain an in-depth understanding of MSRDS. Currently, this book is available on the market (Figure 10.6). Another suggestion is *Professional Microsoft Robotics Developer Studio* (Wrox Programmer to Programmer) (Figure 10.7). From these books, you can obtain more information about the principles of MSRDS and related development skills.

Book name: *LEGO MindStorms NXT X MSRDS*
(in Chinese)
Author: Shih-Chung Kang, Wei-Tze Chang,
Kai-Yuan Gu, and Hung-Lin Chi
Publisher: Delight Press
Publication date: September 2009
ISBN: 9789866348099
Number of pages: 272

FIGURE 10.5   The Chinese edition of this book.

Book name: *Programming Microsoft Robotics Studio*
Author: Sara Morgan
Publisher: Microsoft Press
Publication date: March 2008
ISBN: 0735624321
Number of pages: 288

FIGURE 10.6   Programming Microsoft Robotics Studio.

Book name: *Professional Microsoft Robotics Developer Studio* (Wrox Programmer to Programmer)
Author: Kyle Johns and Trevor Taylor
Publisher: WROX
Publication date: May 2008
ISBN: 0470141077
Number of pages: 826

FIGURE 10.7 Professional microsoft Robotics Developer Studio (Wrox Programmer to Programmer).

Index